PLC Programming & Implementation

An Introduction to PLC Programming Methods and Applications

By

Ojula Technology Innovations

PLC Programming & Implementation

ISBN: 9789493689497

Published in the United States

Table of Contents

About the Author

Bolakale Aremu is a Control Systems Engineer. He has worked with hundreds of clients across the Programmable Logic Controllers (PLC) automation industry. He has written many books and articles, and created courses for higher education institutions. With his 15 years of experience in the Control Systems Engineering, he has helped some world class organizations like Post Holdings, Kraft Heinz and Procter & Gamble.

He has used several PLC-based systems like the Allen-Bradley's RSLogix 5, 500, 5000, Studio 5000 and PACs. He is also a master of other great technologies like Cognex In-Sight Vision Systems to mention a few.

How this Book can Help You

This short book is a valuable introduction to Programmable Logic Controllers, and also a reference guide for beginners. It provides an overview of how relay logic can be converted into PLC logic. I included lots of examples, tables, and ladder diagrams to help simplify the topics I covered for beginners.

You will find this book very helpful if you are an electrician, an instrumentation technician, an automation professional or engineer that wants to improve their knowledge of PLC programming.

Getting Started

Simple Key Terms to Understand Relay Logic to PLC Logic Conversion

Control strategy: This is the sequence of steps that take place in a PLC program, which produces our desired output control.

Control task: It is our desired results of a control program.

Flowcharting: This is a method used to pictorially represent the operation of a process in a sequential manner.

Program coding: This is the process of translating a logic or relay diagram into the PLC ladder program form.

The implementation of a control program requires complex organizational and analytical skills, which change depending on the application. Because they are so varied, we cannot explain how to solve every specific control task.

Nevertheless, we can provide you with techniques and guidelines for completing this problem-solving process. In this handbook, we will introduce a strategy for implementing a control program, which includes program organization, system configuration, and I/O programming.

These strategies also apply to PLCs with the IEC 1131-3 programming standard. Additionally, we will present both simple and complex PLC programming examples. After you finish, you will be ready to learn how to document the PLC system—the last step in implementing the control program.

1. Define the Control Task

The **control task** should be defined by the user to begin the problem-solving process. This is basically defining what must be done. This information is the foundation of the control program. The control task should only be used by people who are familiar with the operation and processes of the machine. The success or failure of the control program directly affects the proper definition of the task.

There are many levels involved in the control task definition process. Each department must collaborate to identify the inputs required so everyone is clear about the project's purpose and scope. If a project involves automation of a manufacturing facility, in which materials are to be retrieved from the warehouse, and then sent to the automated packaging area, both warehouse personnel and those working in packaging areas must work together with the engineering group during system definition.

If the project involves data reporting, management should be involved. The user should look at the steps used to control the task, whether it is done manually or via relay logic. This will help determine if there are any improvements that can be made. Relay logic can be implemented directly in a PLC. However, it is important to redesign the process to meet current project requirements and take advantage of the capabilities of programmable controllers.

2. Determine a Control Strategy

Once the control task is defined, it's time to plan its solution. This involves the creation of a **control strategy**. This is the sequence of actions that must be taken within the program in order to achieve the desired output control. This is the process of developing an algorithm. It may seem strange or new to some, but *algorithm* is not. Everybody uses algorithms to complete certain tasks in their daily lives.

An algorithm is a process that allows a person to move from their home to school or work. The person leaves the house, gets in the car and starts the engine. He or she will reach the destination in the final of a finite number steps.

Implementing a PLC strategy for a control task closely matches the development of an algorithm. The control task must be implemented using a set of basic instructions. It should also be completed in a limited number of steps. If the task of creating an algorithm to solve the problem proves difficult, the user may need to go back to the control task definition in order to redefine the problem.

If we don't know where we are and what Bullfrog County are, then we can't explain how we get there. We need to determine if a specific method of transportation is necessary as part of the problem definition. We need to be aware of any time constraints. Without all the problem definition information, we cannot create a control strategy.

For program strategy definition, the fundamental rule is to think first and program later. You should consider other ways to solve the problem. Before you start programming the control function, it is important to refine the algorithm. This philosophy will reduce programming time reduce the debugging time, speed up start-up and put focus where it is needed - on design when designing and programming when programming.

The system designer must formulate a strategy, no matter if it is a new application of an existing process. The designer must analyze the sequence of events to optimize control by adding or removing steps.

This will require knowledge of PLC-controlled field devices as well as input/output considerations.

3. Implementation Guidelines

Although a programmable controller can be a powerful device, it cannot do more than what is instructed to. The control program is the set of instructions and solution algorithms that the programmer has created for the controller. It gives it all the directions. The organization of the user will determine the success or failure of a PLC control program.

There are many approaches to a problem. However, if an application is managed in a systematic way, there is less chance of making mistakes. The programmer will choose the method that is most appropriate for the program. However, it is important that the programmer follows certain guidelines.

Table1 contains programming guidelines for modernizations and new applications. Modernizations refer to new systems. New applications can be described as systems that are completely new. Modernizations can be used to upgrade existing control systems that were previously inoperative without a PLC (electromagnetic control or individual loop controllers).

New Applications	Modernizations
• Understand the desired function of the system. • Review possible control methods and optimize the process operation. • Flowchart the process operation. • Implement the flowchart by using logic diagrams or relay logic symbology. • Assign real I/O addresses and internal addresses to inputs and outputs. • Translate the logic implementation into PLC coding	• Understand the actual process or machine function. • Review machine logic of operation and optimize when possible. • Assign real I/O and internal addresses to inputs and outputs. • Translate relay ladder diagram into PLC coding.

Table 1. Programming guidelines.

Understanding the machine's operation or process is the first step to solve a control problem. The problem definition should be followed when developing a strategy for new applications. Revision of strategies for new applications and revisions to the control method for modernization projects will help you spot errors made during planning.

The programming stage is where you can see the differences between modernization and new projects. Modernization projects require that the user has a basic understanding of the machine's operation and the task to be performed. A relay ladder diagram like the one in Figure 1 usually defines the sequence in the control program. This ladder diagram can almost be directly translated into PLC ladder charts.

New applications often begin with specifications that are given to the person responsible for designing and installing the control system. These specifications are translated into a written description which explains possible control strategies. To avoid confusion, the written explanation should be clear and concise. This explanation is then used by the designer to create the control program.

4. Program Organization and Implementation

When programming or implementing a control system, organization is key. A larger project will require more organization, especially if there are multiple people involved. A successful control solution is not only about organization. It also requires the ability to implement it.

Programmers must be able to understand the PLC control task, the controlled devices and the equipment required for the job. After these details have been understood, the programmer is able to begin drawing the control program solution.

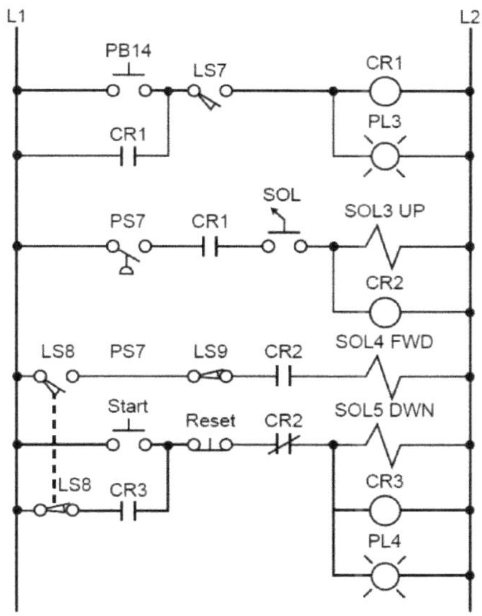

***Figure 1**: Electromechanical relay circuit diagram.*

This time is crucial for project documentation and the system. It can be difficult to document a system after it has been installed and is in use, especially if it doesn't work the first time. Documenting the system from its inception will make it easier to understand the final result.

4.1. How to Create Flowcharts and Output Sequences

Flowcharting is an approach used to plan a program after a detailed written description has been created. A flowchart is an illustration that documents, analyses, and communicates information. It also describes the operational process in a sequence. The flowchart shown in Figure 2 is a simple example. The chart shows a simple flowchart. Each step performs an operation.

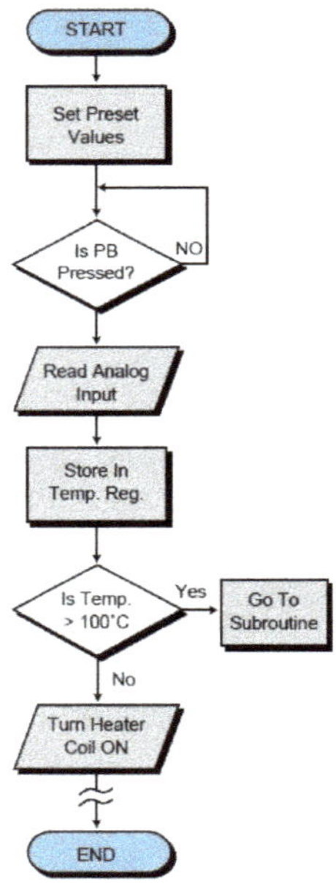

***Figure 2**: Simple flowchart*

A flowchart makes it easy to see the relationship between minor details and broad concepts. A flowchart makes it easy to see relationships and sequences that are difficult to deduce from general descriptions. The flowchart symbols have specific meanings that aid in understanding the solution algorithm. Figure 3 shows the meanings of the most commonly used flowchart symbols.

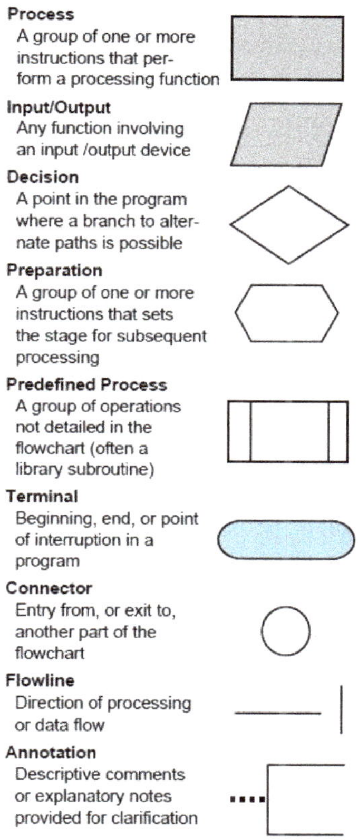

***Figure 3**: Flowchart symbols*

The main flowchart should not be lengthy and complicated. Instead, it should highlight the most important functions that must be performed (e.g. compute engineering units from analogue input counts). To further explain the functions in the main flowchart, you can use several flowcharts.

After the flowchart has been completed, the user may use logic gates or contact symbology in order to implement the logic sequences. PLC contact symbology implements the logic required to program an output step. Logic gates create a logical output sequence based on specific input conditions.

Figure 4 shows both these programming methods. The user should choose the method that is most familiar to them, or a combination of both (see figure 5). However, logic gate diagrams may not be as appropriate for controllers that use Boolean instruction set.

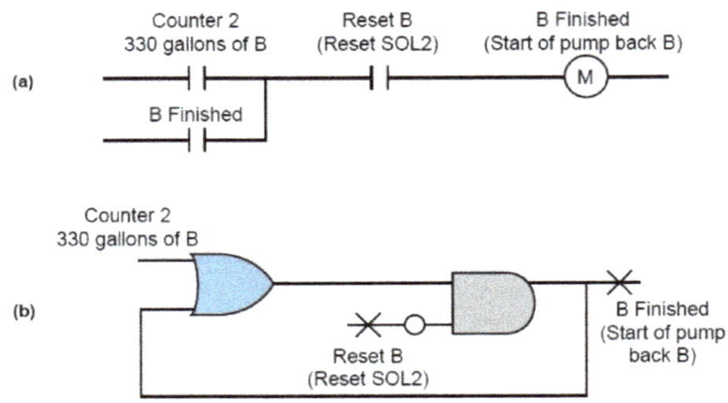

***Figure 4**: (a) PLC contact symbology and (b) logic gate representation of a logic*

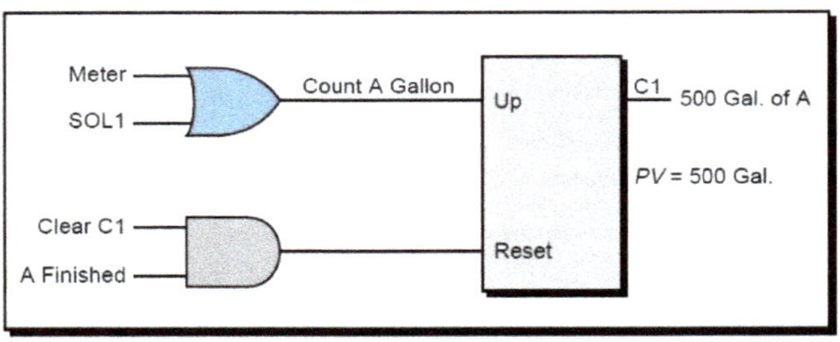

***Figure 5**: A combination of logic gates and contact symbology.*

As shown in Figure 4b (Logic gate diagram), inputs and outputs marked by an X are real I/O. An internal is an I/O point if there is no mark. You can use the actual device names, such as LS1, PB10 or AUTO, to label actual input signals. The labels used for actual input signals can be either the actual device names (e.g. LS1, PB10, AUTO), or symbols letters and numbers that are associated each of the field elements. The user should provide a brief description of the logic sequence during this stage.

4.2. How to Configure the PLC System

During logic sequencing and flowcharting, it is important to consider the configuration of the PLC. Configuration of the PLC determines which I/O modules are used with what types I/O signals. It also specifies where the modules will reside in the rack enclosures. The module locations will determine which I/O addresses will be used by the control program.

When configuring a system, it is important to consider future expansions, special I/O modules such as wire fault inputs or fast-response I/O, and where interfaces will be placed within the rack (all AC/DC I/O together, all low-level analog and DC I/O together). These details should be considered along with the system configuration documentation to ensure a better design.

4.3. Internal and Real I/O Assignment

It is an important step in the implementation and programming organization stages. The I/O assignment tables documents and organizes all of the work done so far. It shows which PLC inputs are connected and which PLC outputs drive what output devices. Here is also where internals are assigned, including counters and timers.

These are the actual coil and contact representations used in the ladder program. If electromechanical relay diagrams exist (e.g. modernization of a process or machine), it is possible to identify real I/O by simply circling the devices, and then assigning I/O addresses (see example 1).

Table 2 contains an I/O assignment table for real inputs or outputs. Table 3 provides an I/O assignment table to internals. These assignments can be extracted directly from logic gate diagrams and ladder symbols used to describe logic sequences. These assignments can also be extracted from the electromechanical diagram's circled elements. The PLC model that was used determines the numbers used to address I/O. These addresses can be represented as octal or decimal. The table's description section lists the field devices that correspond with each address.

Module Type	I/O Address Rack	Group	Terminal	Description
Input	0	0	0	LS1—Position
	0	0	1	LS2—Detect
	0	0	2	Sel Switch—Select 1
	0	0	3	PB1—Start
Output	0	0	4	SOL1
	0	0	5	PL1
	0	0	6	PL2
	0	0	7	Motor M1
Output	0	1	0	SOL2
	0	1	1	PL3

***Table 2**: I/O address assignment table for real inputs and outputs.*

Device	Internal	Description
CR7	1010	CR7 replacement
TDR10	T200	ON-delay timer 12 sec
CR10	1011	CR10 replacement
CR14	1012	CR14 replacement
—	1013	Setup interlock

***Table 3**: I/O address assignment table for internal outputs.*

The input/output connection diagram should be closely followed by the table of address assignments (see Figure 6). Although I/O representations can vary between users due to industry standards, inputs and outgoings are usually represented using squares or diamonds. Part of the documentation package is the I/O connection diagram.

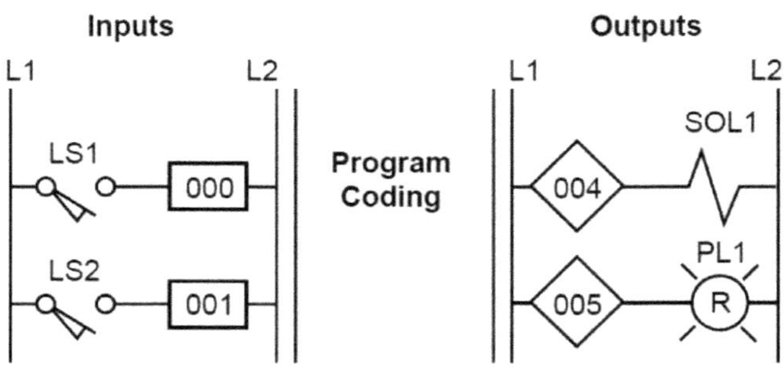

***Figure 6**. Partial connection diagram for the I/O address assignment in Table 2.*

During I/O assignments, users should group the inputs and outputs. This will enable simultaneous monitoring and manipulation of multiple I/O. If 16 motors are to be started in a sequence, they must be grouped together. This will allow monitoring of the I/O registers associated the 16 grouped I/O point will reveal their starting sequence. Because I/O systems are modular, it is important that all inputs and outputs be assigned simultaneously. This will prevent you from assigning an input address to an outgoing module, and vice versa.

4.3.1. Sample Problem 1

Study the circuit shown in Figure 7. **(a)** Identify the real inputs and outputs by circling each, **(b)** Assign the I/O addresses, **(c)** Assign the internal addresses (if required), and then **(d)** draw the I/O connection diagram.

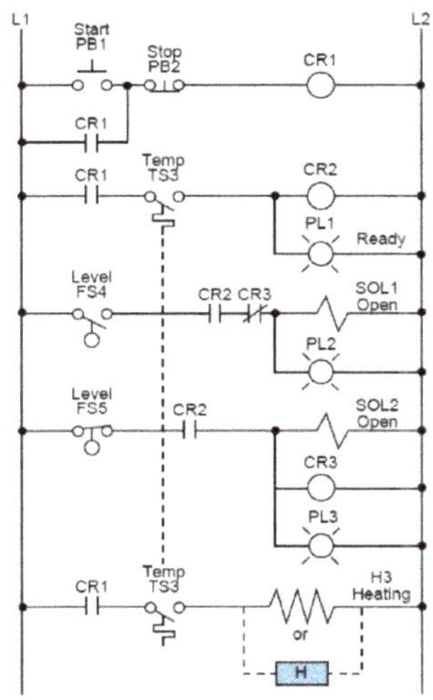

Figure 7: Electromechanical relay circuit.

Assume that the PLC used has a modularity of 8 points per module. Each rack contains 8 modules slots. The master rack is number 0. As long as the right module is used, inputs and outputs can be assigned any address. The PLC determines if an input module or an output module are connected in a slot. Internals begin at address 10008.

4.3.2. Solution to Sample Problem 1

(a) Figure 8 Shows the circled input and output connections. It is important to note that the temperature switch TS3 has been circled twice, even though it is only one device. Only one of the devices is mentioned in the address assignment. Also, only one device is wired to an input unit.

(b) Table 4 illustrates how inputs and outputs are assigned. It assigns all inputs as well as all outputs. There are no I/O locations left for future use.

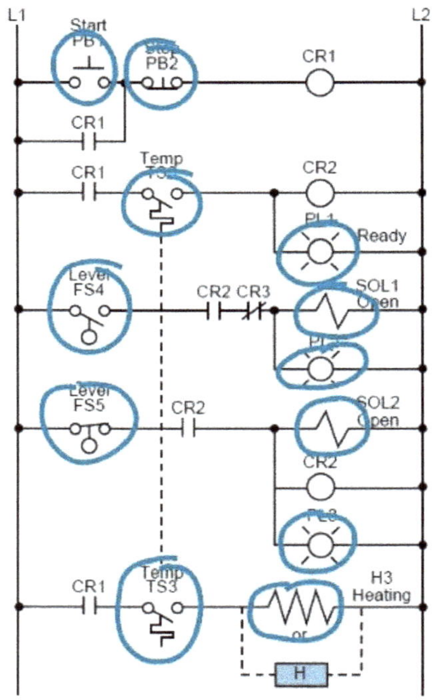

***Figure 8**: Identification of real I/O (circled).*

Module Type	I/O Address Rack	Group	Terminal	Description
Input	0	0	0	Start PB1
	0	0	1	Stop PB2
	0	0	2	Temp TS3
	0	0	3	Level FS4
	0	0	4	Level FS5
	0	0	5	—
	0	0	6	—
	0	0	7	—
Spare	0	1	0	Not used
	:	:	:	
	0	1	7	
Output	0	2	0	PL1 Ready
	0	2	1	SOL1 Open
	0	2	2	PL2
	0	2	3	SOL2 Open
	0	2	4	PL3
	0	2	5	H3 Heating
	0	2	6	—
	0	2	7	—

***Table 4**: I/O address assignment.*

(c) Table 5 presents the output assignments and a description for each internal. It is important to note that the control relay CR2 does not count as an internal as it is identical with the output rung for PL1. Every contact associated with CR2 is replaced by 020 (the address for PL1) when the control program has been implemented.

Device	Internal	Description
CR1	1000	Control relay CR1
CR2	—	Same as PL1 Ready
CR3	—	Same as SOL2 Open

***Table 5**: Internal output assignment.*

(d) Figure 9 shows the I/O connection diagram of the circuit in Figure 7. This diagram is based upon the I/O assignment in part (b). Only one of the temperature switches (the normally open TS3 switch) is a connected input. Each switch's logic programming should be based upon a normally open condition.

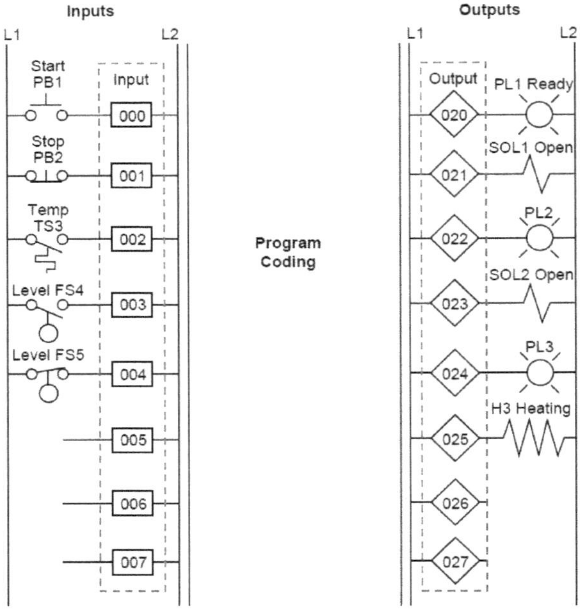

***Figure 9**: I/O connection diagram.*

4.4. Assignment of Register Addresses

Another important aspect of PLC organization is the assignment of addresses to registers used for control programs. It is easiest to list all available PLC registers in order to assign them. Next, you will need to describe the contents, function, and description of each register in a register assignment table. The register assignment table shown in Table 6 for the 15 first registers of a PLC system ranges from address 20008 up to address 20168.

Register	Contents	Description
2000	Analog input	Temperature input temp 3 (inside)
2001	Analog input	Temperature input temp 4 (outside)
2002	Spare	–
2003	Spare	–
2004	TWS input	Set point (SP1) input from TWS panel 1
2005	TWS input	Set point volume (V1) from TWS panel 2
2006	Constant 2350	Timer constant of 23.5 sec (0.01 sec TB)
2007	Accumulated	Accumulated value for counter R2010
2010	Spare	–
2011	Spare	–
2012	Constant 1000	Beginning of look-up table (value #1)
2013	Constant 1010	Look-up value #2
2014	Constant 1023	Look-up value #3
2015	Constant 1089	Look-up value #4
2016	Constant 1100	Look-up value #5

***Table 6**: Register assignment table*

4.5. Elements Not to Be Wired to the Controller

The user must decide which devices will be disconnected from the controller when assigning inputs and outputs. These elements wil not be wired to the controller (electromagnetic control logic). These elements often include devices such as hydraulic pumps and compressors that are rarely turned off after start. For safety reasons, components such as master start push buttons and emergency stops should be hardwired. If the controller fails or the system becomes unresponsive, the user can turn off the system by simply pressing the button.

Figure 10 shows an example of system components that are often left hardwired. The normally closed PLC Fault Contact 1 or watchdog timer contacts is wired in series to other emergency conditions. The contact is closed when the controller operates correctly but opens when there is a fault. This contact can be used by the system designer to shut down the PLC system in an emergency.

PLC fault contacts can be used to improve or implement a safety circuit. If a PLC operates correctly, the normally opened fault contact will close and the normally closed one will open when the PLC turns on. These contacts are connected in series to the hardwired circuit as shown in Figure 10. This means that, if the PLC stops working during normal operation, the normally opened contacts will also open. The hardwired circuit will be shut down at the point that the PLC is the controlling element. To control the power supply to the other control components, this circuit also uses an SCR safety control relay (SCR). To indicate an alarm condition, the normally closed fault contacts can be used.

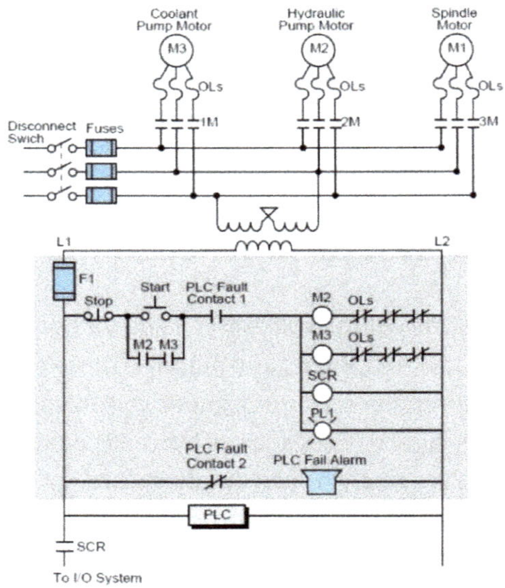

***Figure 10**. Hardwired components in a PLC system.*

Figure 10 shows how an emergency situation, such as a malfunctioning PLC, will cause power to be cut (L1) at the I/O module. Turning OFF the safety relay (SCR), will cause the SCR to open, preventing power from flowing to the system. The normally closed PLC fault contact (2PLC Fault Contact 2) will notify personnel of system failures due to a PLC malfunction. This type of alarm should be implemented by the designer in the main PLC rack as well as each remote I/O rack. Remote systems also have fault contacts integrated into their remote controllers. This allows subsystem problems to be notified quickly so that they can be addressed without putting people at risk.

4.6. Special Input Device Programming

Special programming is required for certain PLC circuits or input connections. Programming normally closed input devices is one example. It is important to remember that how a device behaves in a control program affects its programming.

Normally Closed Devices A normally closed input device can be programmed so that it acts as either a normally opened or normally closed device. For normally closed inputs, the same rules apply. The rule is that if a device wired as a normally close input and must act as such, its reference address will be programmed as normally opened. The following example shows that a normally closed device within a hardwired circuit will be programmed as normally open when it is replaced by the PLC control program. The program doesn't evaluate the device as an input because it isn't referenced as such.

4.6.1. Sample Problem 2

Draw the PLC ladder program for the circuit shown in Figure 11 and create an I/O address assignment list. Use addresses 10_8 to 47_8 for inputs. Start outputs at 50, and internals at 100_8.

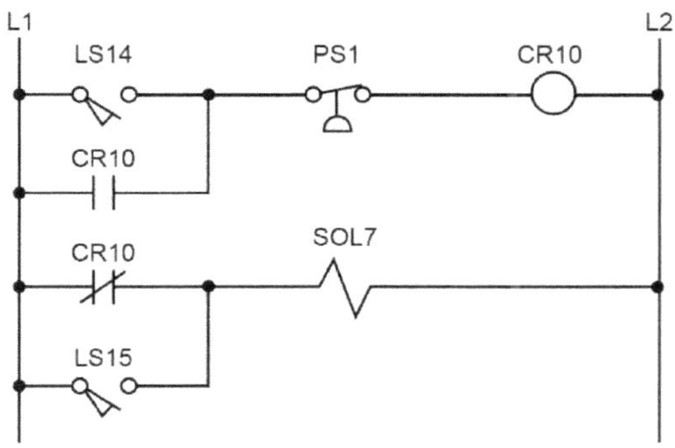

Figure 11: Electromechanical relay circuit.

4.6.2. Solution to Sample Problem 2

Figure 12 shows an equivalent PLC ladder diagram of the circuit shown in Figure 11. The I/O address assignment table is shown in Table 7. Because internal coil 100 refers to it, the normally closed contact (CR10) must be programmed as such.

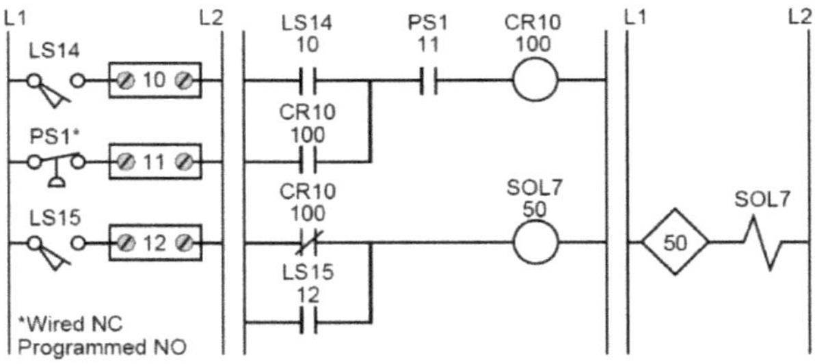

Figure 12: PLC ladder diagram of the circuit in Figure 11.

I/O Address	Device	Type
10	LS14	Input
11	PS1	Input
12	LS15	Input
50	SOL7	Output
100	CR10	Internal

***Table 7.** I/O address assignment table.*

Master Control Relays. A master control relay (MCR) is another circuit that programmers should be aware. An MCR coil is used to control several rungs of an electromechanical circuit diagram by switching ON/OFF power to the rungs. An MCR coil is not able to control the end of a hardwired circuit unless the circuit is fully followed. In Figure 13, for example, the MCR output at line 1 controls power to hardwired elements starting from line 3, which is where the MCR contact is, and ending at line 51.

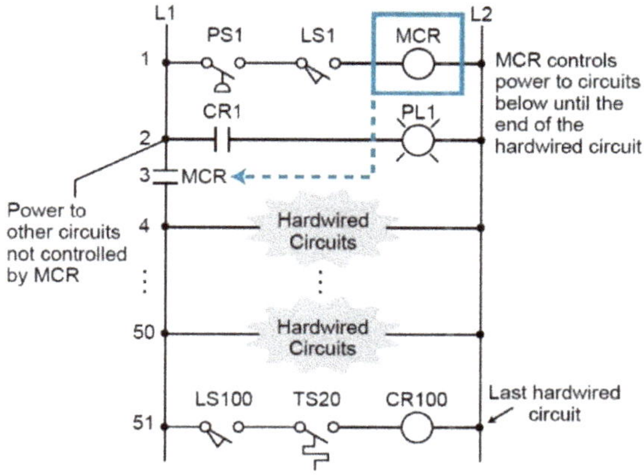

***Figure 13**: Electromechanical relay circuit with a master control relay.*

When the master control relay (ON), power will flow to these lines (4 through 51). If the master control relay turns off, power will not flow to these rungs (lines 4 through 51). These devices will not be able to execute the control action. This configuration is similar to a hardwired subprogram, or subroutine. If the master control relay is ON, the rungs will be executed. If it is OFF the rungs will not be executed. The circuit's line 2 contains power branches that connect to other circuits not affected by MCR's actions. These circuits are the standard hardwired program.

The programmer must place the END MCR instruction following the last rung that the MCR should control during the translation from a hardwired ladder to PLC symbology. Figure 14 shows the location of the MCR instruction in the circuit. Figure 14 shows the placement of the MCR instruction for the circuit in Figure 13. This is how the hardwired circuit works.

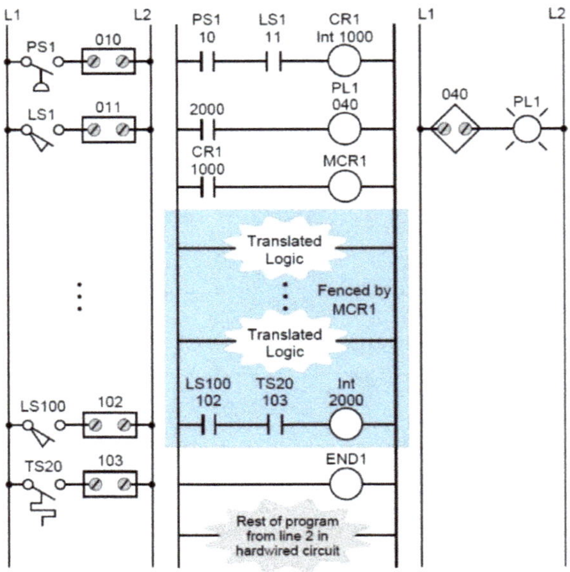

Figure 14: PLC ladder diagram with MCR fence.

The MCR fence is ended by the END1 instruction. After the END1 instruction, the instructions corresponding to hardwired circuits which branch from line 2 of the electromechanical diagram in Figure 13 are found. Figure15 shows a partial ladder of an elaborate circuit with this MCR type. After the rung containing PL3 output, the corresponding PLC program must have an ENDMCR.

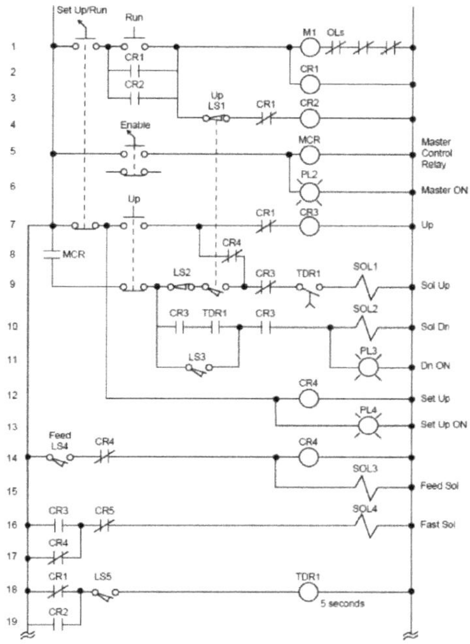

***Figure 15**: Electromechanical relay circuit with an MCR.*

4.6.3. Sample Problem 3

In Figure 15, highlight the sections that will be controlled by a PLC MCR. What additional steps must be taken in order to bypass or include other hardwired circuits inside the MCR fence

4.6.4. Solution to Sample Problem 3.

Figure 16 shows the circuits that need to be fenced according to the MCR instruction. The MCR fencing does not include solenoid SOL1 or part of its driving logic. However, SOL1, CR3, TDR1 and TDR1 may be turned ON using logic before the MCR fence (see Figure 17.). The MCR fence must be properly programmed by the PLC program. It must contain two internal control relays to take out SOL1.

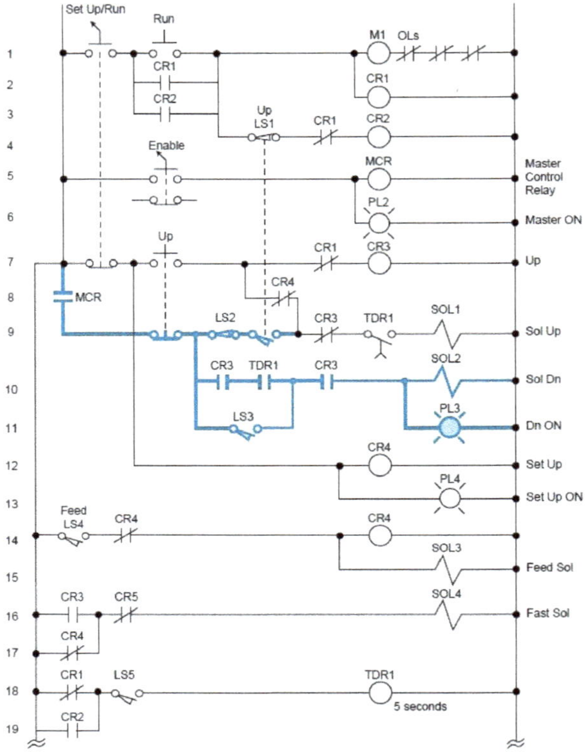

***Figure 16**: MCR-controlled program elements.*

Figure 18 shows the fenced circuit with additional internals (CR1000 & CR1001). The instructions in this diagram are the same as the hardwired circuit. Because the solenoid SOL1 can be turned ON or off by either the outside logic (highlighted in Figure 17), or the logic within the MCR fencing (highlighted in Figure 18), it will be located outside the MCR gate.

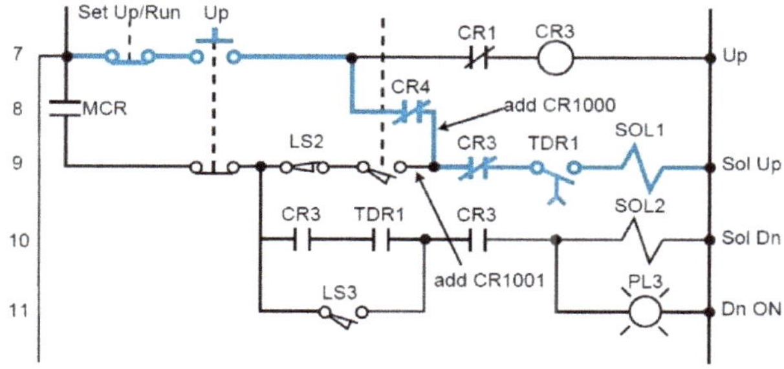

Figure 17: *SOL1 activated by logic outside of the MCR fence.*

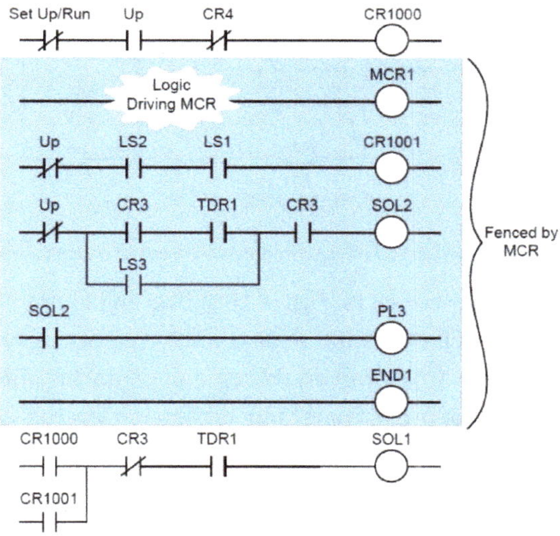

Figure 18: *MCR fence.*

Bidirectional Power Flow. Figure 19 shows another problem that could cause programming problems: bidirectional power flow through the normally sealed CR4 contact in line 8. The bidirectional flow problem can be solved by knowing whether CR4 has an effect on the output rungs it is connected to. These rungs are the solenoid SOLO1 output and the control relay output CR3. Figure 19 shows the two possible paths in a hardwired circuit. PLCs allow only forward paths. Therefore, if the circuit's logic requires a reverse path, the logic driving the output CR3 must include the CR4 contact (see Figure 9b).

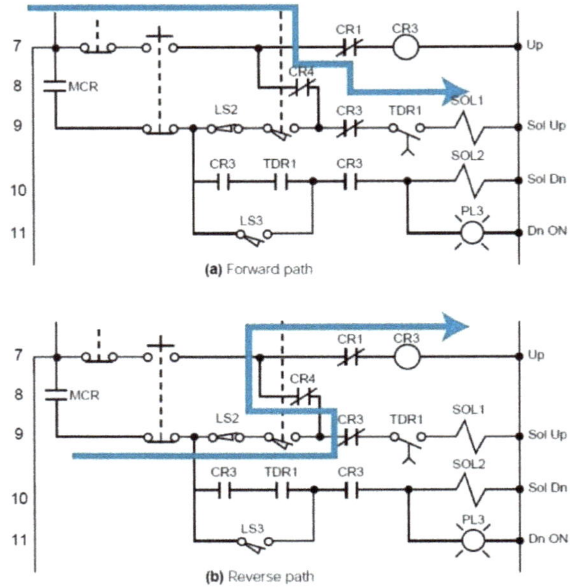

***Figure 19**: (a) Forward and (b) reverse power flow in a hardwired circuit.*

Instantaneous timer contacts. Figure 15 shows an electromechanical circuit that specifies an instantaneous contact (the normally closed TDR1 contact in line 10) However, this type of contact is not usually available in PLCs. The programmer must create an internal output to capture the timer and then use the internal contact to drive the logic of the timer.

Figure 20a shows an electromechanical circuit. If PB1 or LS1 close simultaneously, the timer will begin timing, and the instantaneous contacts (TMR1-1-1) will close. This seals PB1. The circuit is sealed if PB1 is removed (OFF). Figure 20b shows how to trap a timer. An internal output of the PLC program traps the instantaneous contact coming from the circuit's electronic timer. The contacts are controlled by the internal drive of the timer. If a trap is not present, the timer will begin timing when PB1 or LS1 close. However, it will stop timing once PB1 is released.

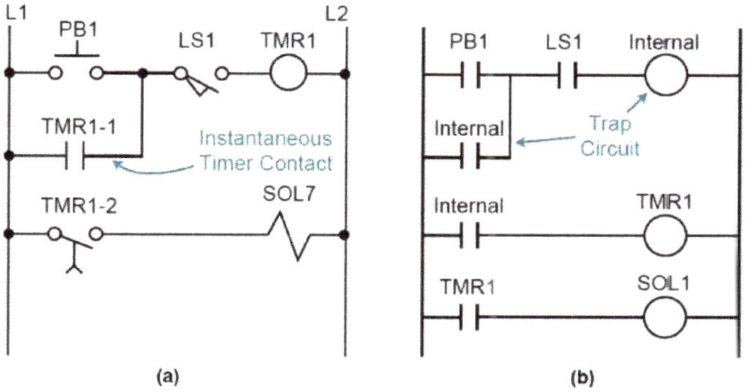

***Figure 20**: (a) An instantaneous timer contact in a hardwired circuit ana (b) a trapped*

Complex Logic Rungs. The best programming technique for a confusing logic rung is to isolate it. Next, you will need to reconstruct all possible logic paths starting at the output and ending where the rung begins. It may be easier to create an inner output at the intersection of two rungs if a section of a rung, such as the one in Example 3, is directly connected or interacts with another. Next, you can use the internal output for driving the rest of your logic. This cross point is located in Line 9 of the Figure 15 circuit. It is between the normally open LS1 (or normally closed CR3).

4.7. Program Coding/Translation

Program Coding is the process of converting a logic diagram or relay diagram into a PLC ladder program form. The ladder program is what will control the machine or process. It is stored in the application RAM. The order of the previous stages (control task description, I/O assignment etc.) directly affects how easy it is to code programs. a program code has been completed. Figure 21 shows an example program code that was generated from electromechanical relay diagrams and logic gates (internal coil 1,000 replaces the control relay). The coding represents the logic in a PLC format. It doesn't matter if it's a new application, or modernization. This section will provide several programming examples and a closer look at the coding process.

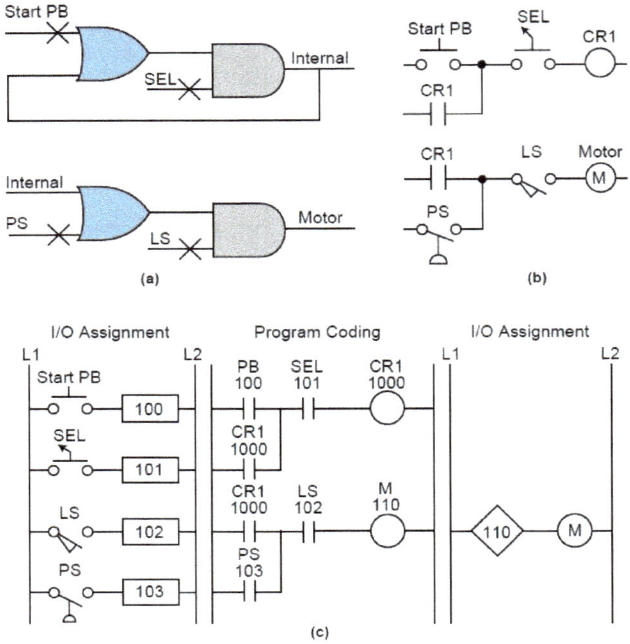

***Figure 21**: Translation from (a) logic gates and (b) an electromechanical relay diagram.*

5. Discrete I/O Control Programming

This section will provide programming examples to illustrate modernization of relay system. Examples will be provided that relate to the new PLC control implementations. These examples will mainly focus on discrete controls. Next section will discuss programming and analog I/O interaction.

5.1. Control Programming and PLC Descriptions

Modernization is the process of moving a machine's or process's control over from traditional relay logic to a programmable controller. The control logic is housed in traditional hardwired relay panels. They can be difficult to maintain, including contact chatter and contact welding. The machine's performance can be improved by switching to a PLC. It also allows for better control. Actually, the "new" program for programming the machine's controller is based on the original hardwired system's instructions and controls.

This section will show you how to set up and configure a PLC program. Each module has four I/O points in the controller's I/O structure. There are eight racks (0-7), each with eight slots or groups where modules can be placed. This configuration is illustrated in Figure 22.

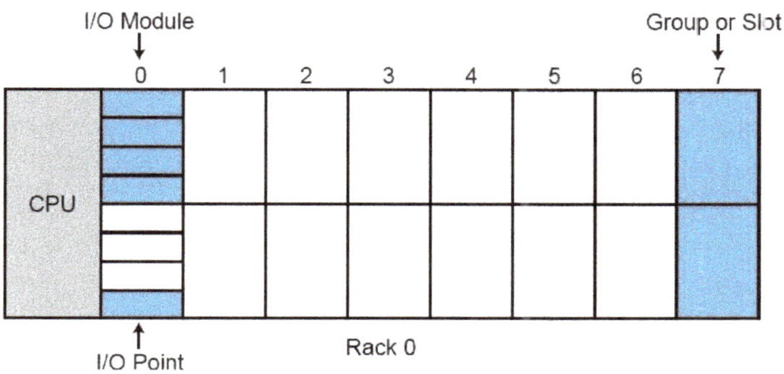

***Figure 22**: Example PLC configuration.*

The PLC can handle four-channel analog input modules. These can be placed anywhere in the slot. If analog I/O modules have been used, they cannot be placed in the same slot as discrete I/O. Multiplexed registers can be accepted by the PLC. Multiplexed modules have two slots and provide enable (select), lines for I/O devices.

Addresses 000 to 777 octal are input and output device connections that have been mapped to I/O tables. The address's first digit is the rack number. The second digit is the slot and the third digit the terminal connection. The PLC determines whether the slot has an input or output.

Point addresses 1000_8 to 2777_8 can be used as internal outputs. Register storage begins at register 3000_8 and ends with register 4777_8. There are two types of counter and timer formats: ladder format and block format. However, all timers need an internal output to indicate the ON-delay output. Block format timers put a "T" in front of their internal output address. Ladder format timers add a "T", while block format timers specify that address in the block's output coil.

For discrete inputs, addresses 000_8 through 027_8 will be used in the following sections. Addresses 030_8 through 047_8 will be used for outputs. When possible, analog I/O will go in the last slot on the master rack (0). These examples will show you that sometimes, the assignment of registers and internals can be done simultaneously with the programming stages.

5.2. Simple Relay Replacement

This relay replacement example uses the PLC implementation for the electromechanical circuit in Figure 23.

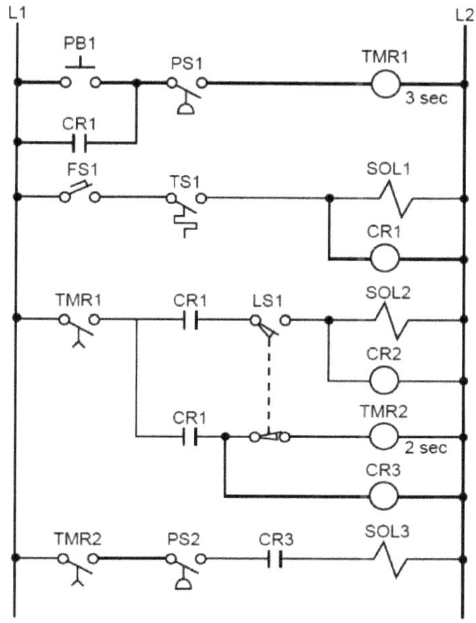

***Figure 23**: Electromechanical relay circuit.*

Hardware timer TMR1 requires instantaneous contacts in the first rung, which are used for latching the rung. If instantaneous TMR1 contacts were implemented using a PLC timedelay contact, then PB1 should be pushed until the timer has set to latch the rung. This instantaneous contact is implemented by trapping a timer with an external output.

Tables 8 and 9 give the I/O address as well as the internal output assignments for an electromechanical circuit's true I/O. The register assignment table is shown in Table 10. The internals cannot replace the control relays CR1 or CR2, since output addresses 030 and 031 that correspond to solenoids (SOL1 and SOL2) are available.

Addresses 030 or 031 can be used to replace the CR1 or CR2 contacts wherever they appear in the program. The normally open contact LS1 connects to the PLC input interface. The normally open LS1 refers to the normal closed LS1 program. The PLC program coding solution is illustrated in Figure 24.

Module Type	I/O Address Rack	Group	Terminal	Description
Input	0	0	0	PB1
	0	0	1	PS1
	0	0	2	FS1
	0	0	3	TS1
Input	0	0	4	LS1
	0	0	5	PS2
	0	0	6	—
	0	0	7	—
Output	0	3	0	SOL1
	0	3	1	SOL2
	0	3	2	SOL3
	0	3	3	—

Table 8. I/O address assignment.

Device	Internal	Description
TMR1	1000	Used to trap TMR1
CR1	—	Same as SOL1 (030)
CR2	—	Same as SOL2 (031)
TMR1	1001	Timer TMR1
TMR2	1002	Timer TMR2
CR3	1003	Replace CR3

Table 9. Internal address assignment.

Register	Description
4000	Preset timer count for 3 sec
4001	Accumulated count timer 1001
4002	Preset timer count for 2 sec
4003	Accumulated count timer 1002

Table 10. Register assignment.

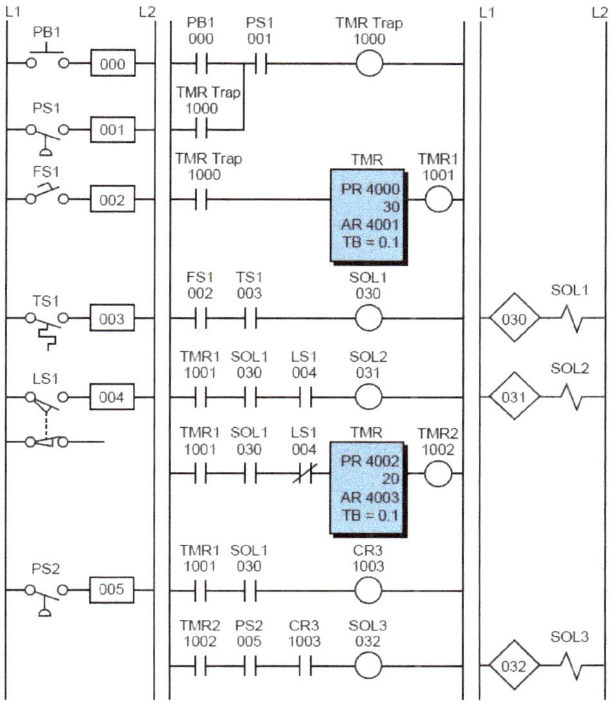

Figure 24: *PLC implementation of the circuit in Figure 23.*

5.3. Simple Start/Stop Motor Circuit

Figure 25 shows the wiring diagram of a three-phase motor and its three-wire control circuit. The starter push button is sealed by the auxiliary contacts. This circuit can be converted into a PLC program by first identifying which control devices will make up the PLC I/O systems. These are the items circled in Figure 26. The circuit will include the inputs and outputs for the start and stop buttons and the starter coil. Because the starter coil's auxiliary contacts are not part of the system, an internal will seal it. This will result in less wiring and fewer connections.

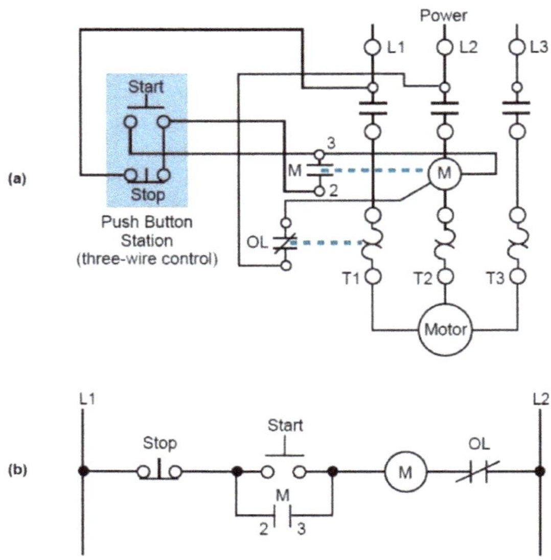

Figure 25: *(a) Wiring diagram and (b) relay control circuit for a three-phase motor.*

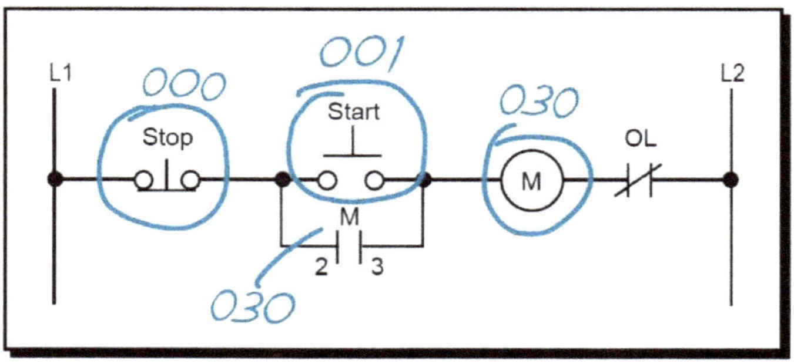

Figure 26: *Real inputs and outputs to the PLC.*

Table 11 lists the I/O address assignment. It uses the same addressing scheme that the circuit diagram uses (that is, inputs: addresses 000, 001, output: address 030).

The PLC must be programmed in the exact same logic sequence as the hardwired circuit. (See Figure 27). The stop push button will therefore be programmed in parallel with the start button as an examineON instruction.

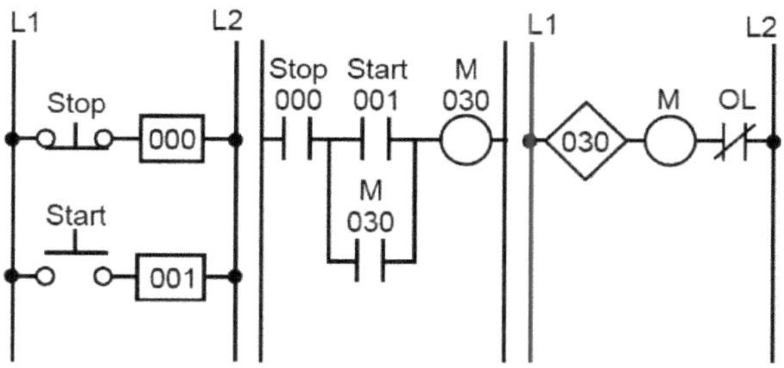

__Figure 27__: PLC implementation of the circuit in Figure 25.

This circuit will drive output 0, which controls the starter. When the start button is pressed output 030 will turn on, sealing the start button and turning the starter ON. The motor will turn off if the stop button is pressed. The stop push button can be wired as normal to the input module. The starter coil's overloads can be wired in series to the coil.

A PLC wiring diagram shows the PLC connected to the power lines L1 (see Figure 28). The module's field inputs are connected to L1 and the module to the other. The return connection to the input module is L2. L1 is where the output module receives power to switch the load. The output terminal 030 is connected to the starter coil and its overloads. They go to L2.

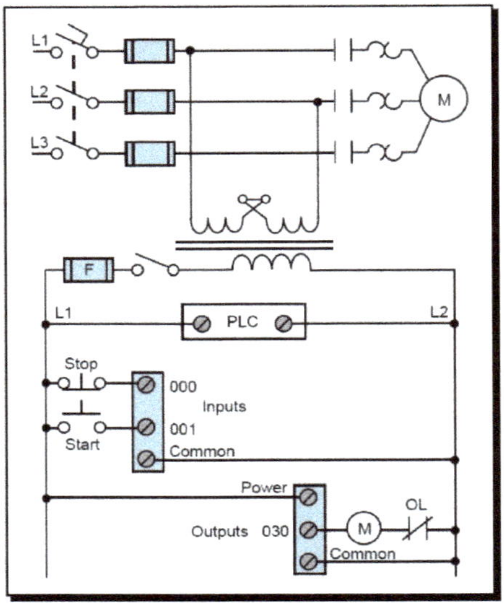

Figure 28: PLC wiring diagram of a three-phase motor.

For proper operation, the output module can also be connected directly to L2. The wiring diagram for the motor control circuit (see Figure 29) shows that the PLC output module connects directly to the starter coil.

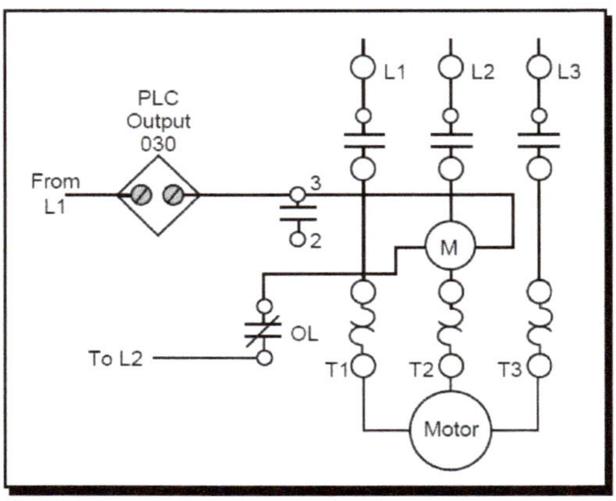

Figure 29: Motor control circuit's wiring diagram.

Even though the 3-phase motor has a 3-wire control circuit, its corresponding PLC control circuit has only 2 wires. This 2-wire configuration is similar to a 3-wire configuration because it provides low-voltage release; but it does not provide low-voltage protection. In Figure 29, the starter's seal-in contacts (which is labeled as 3—| |—2) are shown as unconnected because they are not used. If the overload opens while the motor is running, the motor will stop, but the circuit will remain ON. Immediately the overloads cool off, and the overload contacts close, the motor will start again at once.

This situation could be undesirable, depending on the application. This could be a problem with a motor, or someone might be trying to fix it. The motor could then suddenly start up again. This can be avoided by making the auxiliary contact an input, and using its address as the start button seal. (See Figure 30). This configuration means that if overloads are opened while the motor is running the coil will turn off, and their auxiliary contacts will interrupt the circuit in PLC.

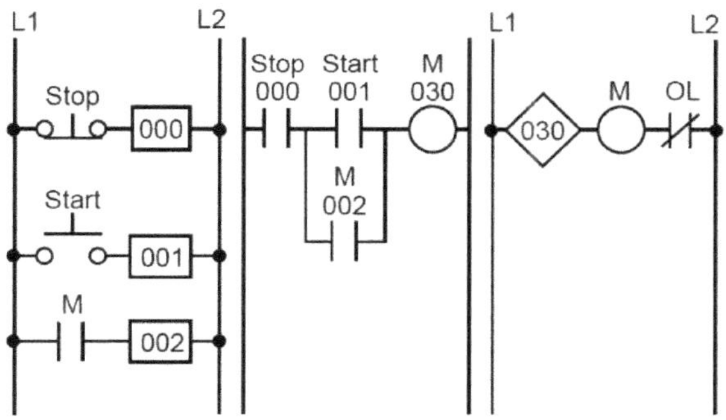

***Figure 30**: Two-wire circuit configured as a three-wire circuit*

5.4. Forward/Reverse Motor Interlocking

Figure 31 shows a hardwired forward/reverse circuit with push button and electrical interlockings. The simplified wiring diagram of this motor is shown in Figure 32.

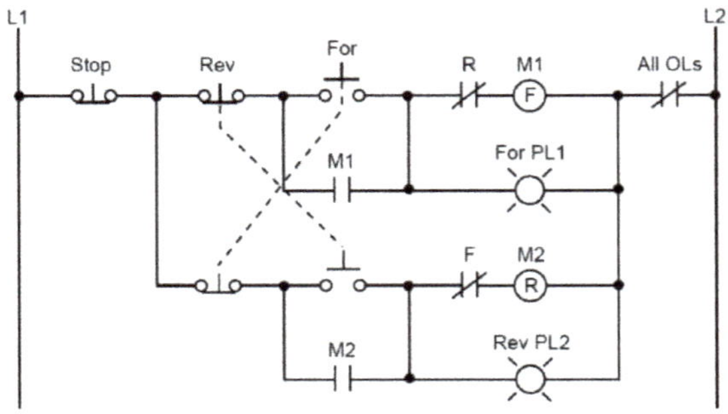

Figure 31*: Hardwired forward/reverse motor circuit.*

This circuit must be implemented using the overload contacts in the PLC. They are used to monitor the existence of overload conditions. Auxiliary starter contacts M1 and M2 are not necessary in the PLC program. Sealing circuits can be programmed using internal contacts from motor outputs. The overload contact input can be used to provide low-voltage protection so that the motor circuit will shut down in the event of overload. To restart the motor, however, the operator must press the forward or reverse button once the overload condition has passed.

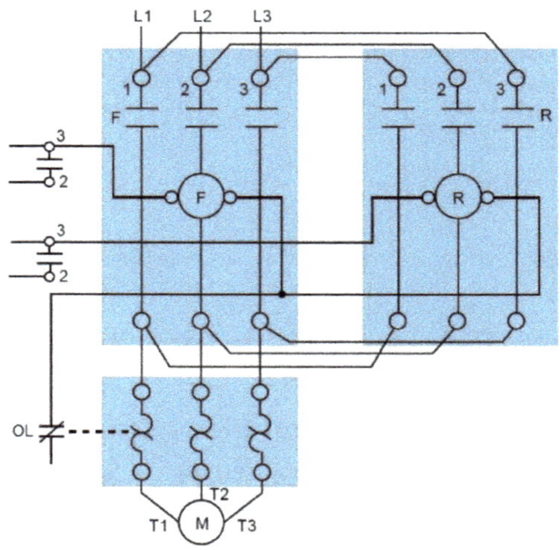

Figure 32*: Forward/reverse motor wiring diagram.*

The circuit shown in Figure 31 is a PLC implementation. It does not include the starter contacts R and F, which are normally closed in the hardwired circuit. However, push button interlocking does the same job. This redundant interlock acts as an additional interlocking mechanism in the hardwired circuit.

Figure 33 illustrates the field devices that will connect to the PLC. Address 000 is the stop button, while addresses 001 and 022 are the normally open sides for the forward and reverse buttons.

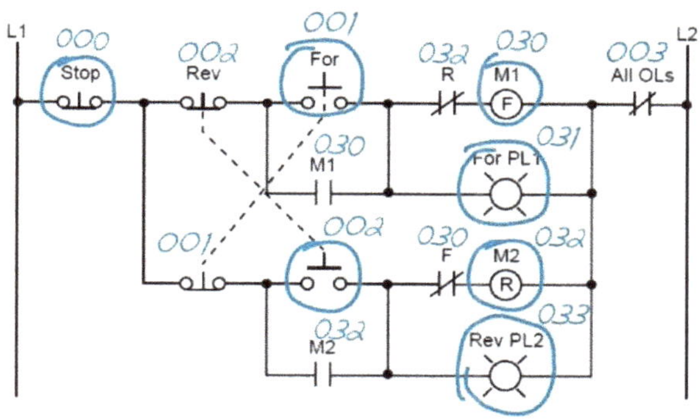

Figure 33: *Real inputs and outputs to the PLC.*

At address 003, the overload contacts are connected with the input module. Addresses 030 and 032 are the output devices, which include forward and reverse starters as well as their interlocking auxiliary contacts. The forward and reverse pilot lights have addresses 031 and 033, respectively. The overload light indicators also have addresses 034 or 035, which indicate that an overload condition occurred while either the forward or reverse motor were operating.

Addresses for the auxiliary contacts interlocking using R and F contacts are output addresses of forward and reverse starters (030, 032). As we will see, the ladder circuit that latches an overload condition (forward/reverse) must be programmed prior to the circuits driving the forward or reverse starters (030 and 032). The overload signal will not be recognized by the PLC program if the starter is turned off in the circuit at the same time as the overload. The latching circuit must be placed after the motor starter circuit. This is because the starter contacts are open and there will not be continuity.

Table 12 shows the actual I/O address assignment of this circuit. Figure 34 illustrates the PLC implementation. It follows the same logic and has additional overload contact interlockings. The overload input will also be used by the motor circuit, which will turn off the motor. Programming the motor starter outputs' logic to open the normally closed overload contacts is done as normal.

Module Type	I/O Address Rack	Group	Terminal	Description
Input	0	0	0	Stop PB (wired NC)
	0	0	1	Forward PB (wired NO)
	0	0	2	Reverse PB (wired NO)
	0	0	3	Overload contacts
Input	0	0	4	Acknowledge OL/Reset PB
	•	•	•	
	•	•	•	
	•	•	•	
Output	0	3	0	Motor starter M1 (FWD)
	0	3	1	Forward PL1
	0	3	2	Motor starter M2 (REV)
	0	3	3	Reverse PL2
Output	0	3	4	Overload condition FWD
	0	3	5	Overload condition REV
	0	3	6	—
	0	3	7	—

Table 12. I/O address assignment.

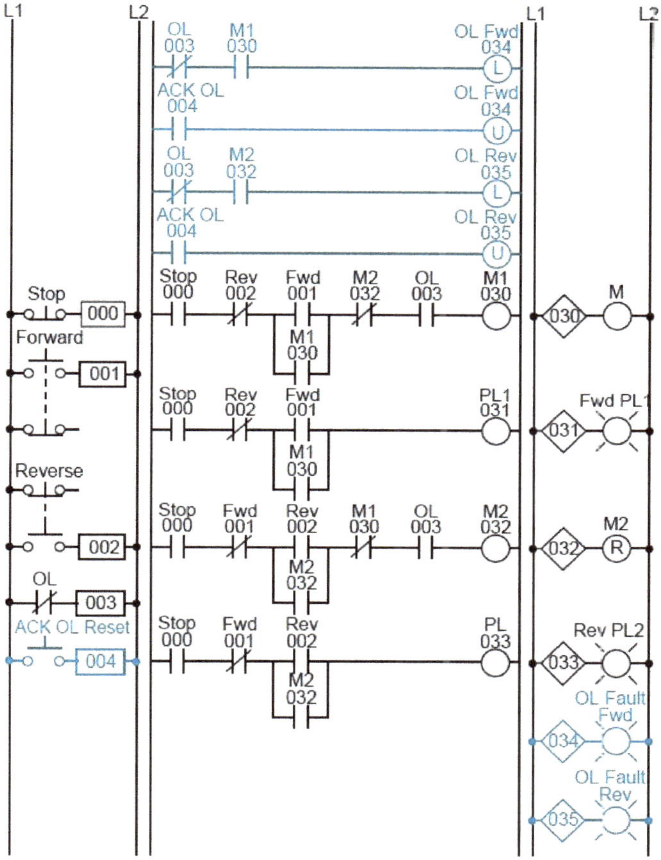

***Figure 34**: PLC implementation of the circuit in Figure 31.*

If there is no overload condition, the forward and reverse motor commands will work normally. The overload contacts will ensure continuity. If overload occurs, however, the contacts in PLC program will be opened and the motor circuit will turn off. The overload indicator pilot light (OL Fault Fwd & OL Fault Rev) uses latch/unlatch instructions for latching whether overload occurred in forward or reverse.

The latching takes place before the forward and reverse motor starter circuits. These will shut off due to overload. The operator can reset the overload indicators by pressing the normal open acknowledge overload reset button that is attached to the input module. The overload indicators will still be locked, even after the physical overloads cool down and return to normal closed states until the operator acknowledges that the condition has been rectified and resets them.

Figure 35 shows the motor wiring diagram for the forward/reverse circuit and the output connections to the PLC. The auxiliary contacts M1 & M2 are not connected. The wiring diagram shows that both the forward coil and reverse coils have the returns connected to L2 but not the overload contacts. The overload contacts are connected on one side to L1 and the PLC's input (input 033) on the other. The overload condition will cause both motor starter output coils to be removed from the circuit. This is because both starters are connected to the PLC's input module (input 003).

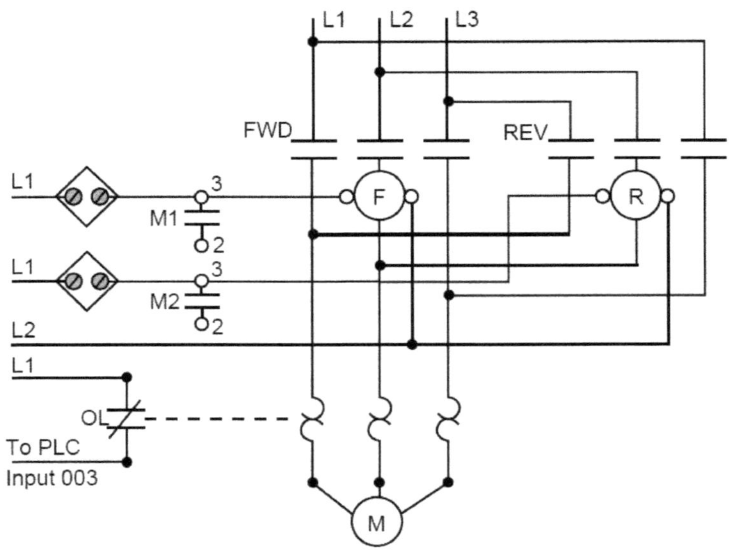

***Figure 35**: Forward/reverse motor wiring diagram.*

5.5. Reduced-Voltage-Start Motor Control

Figure 36 illustrates the control circuit and wiring diagram of a 65% tapped, autotransformer, reduced-voltage-start motor control circuit. The reduced-voltage start reduces the motor's inrush current (locked-rotor current), to only 42% at full speed. The timer should be set at 5.3 seconds in this example. The instantaneous contacts in lines 2 and 3, must also be captured.

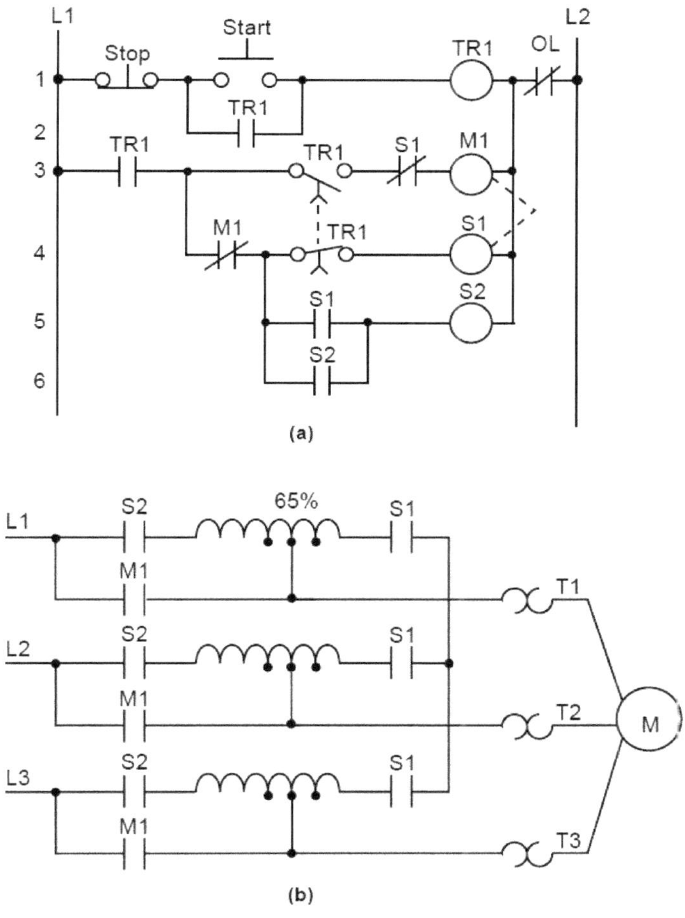

Figure 36: (a) Hardwired relay circuit and (b) wiring diagram of a reduced-

voltage-start

Figure 37 shows the hardwired circuit, with the actual inputs and outputs highlighted. Devices that aren't circled can be implemented within the PLC by programming internal instructions. The I/O assignment and internal assignment are shown in Tables 13-14 and 15, respectively.

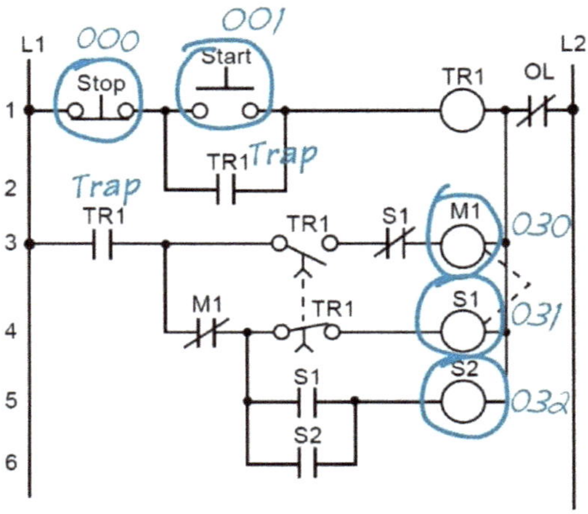

***Figure 37**: Real inputs and outputs to the PLC.*

Figure 38 illustrates the PLC implementation of the reduced-voltage-start circuit. The first line of the PLC program traps the timer with internal output 1000. Contacts from this internal replaces the instantaneous control circuit timer contacts.

Module Type	I/O Address			Description
	Rack	Group	Terminal	
Input	0 0	0 0	0 1	Stop PB (NC) Start PB (NO)
Output	0 0 0	3 3 3	0 1 2	Motor Starter M1 S1 S2

Table 13. I/O address assignment.

Device	Internal	Description
—	1000	Trap timer circuit
Timer	1001	Timer

Table 14. Internal address assignment.

Register	Description
4000	Preset register value 53, time base 0.1 sec for 5.3 sec (timer output is 1001)
4001	Accumulated register for timer output 1001

Table 15. Register assignment.

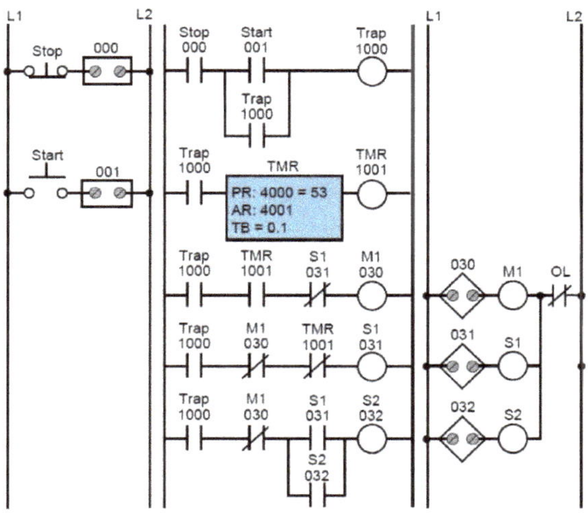

Figure 38: *PLC implementation of the circuit in Figure 36.*

The PLC circuit implementation doesn't provide low-voltage protection because the interlocking uses the physical inputs M1, S1 and S2. Low-voltage protection can be provided by programming the starter's overload contacts or auxiliary contacts. If overloads or auxiliary contacts are being used as inputs they must be programmed normally open (closed when overloads are closed and motor is running) then placed in series with contact 1000 at line 3 of the PLC software. The circuit will lose continuity if the overloads are open and M1 will turn off.

5.6. AC Motor Drive Interface

Speed control of AC motors with variable speeds (VS) drives is a common PLC application. Figure 39 illustrates an operator station that can be used to control a VS motor manually. This station's programmable controller implementation will allow for automatic motor speed control via an analog interface. The output voltage can be varied (0-10 VDC) to the drive.

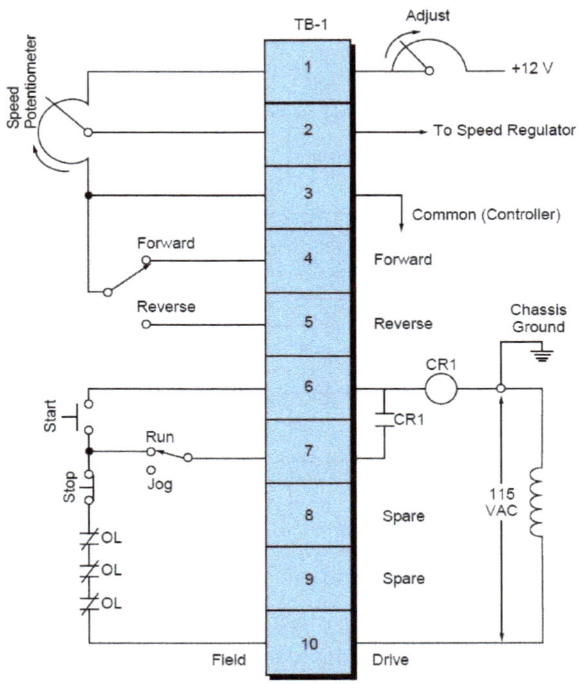

Figure 39: Operator station for a variable speed drive.

The operator station is composed of a speed regulator, forward/reverse direction selector and start/stop push buttons. All of these inputs will be included in the PLC program, except for the potentiometer which will be replaced with an analog output. The application will include the required input field devices, such as the start push button and stop push buttons, jog/run and forward/reverse. These inputs will be connected to input modules and not used by the operator station's parts. The logic to initiate, stop, and interlock forward/reverse commands will be contained in the PLC program.

Table 16 lists the I/O address assignment tables for this example. Figure 40 shows the connection diagram between the PLC and the VS drive's terminal blocks (TB-1).

	I/O Address			
Module Type	**Rack**	**Group**	**Terminal**	**Description**
Input	0 0 0 0	0 0 0 0	0 1 2 3	Start Stop Forward/reverse selector Run/jog selector
Output 115 VAC	0 0 0 0	3 3 3 3	0 1 2 3	Drive enable (L1 from drive)
Output Contact	0 0 0 0	3 3 3 3	4 5 6 7	Forward Reverse
Analog Output	0 0 0 0	7 7 7 7	0 1 2 3	Analog speed reference 0–10 VDC

Table 16. I/O address assignment.

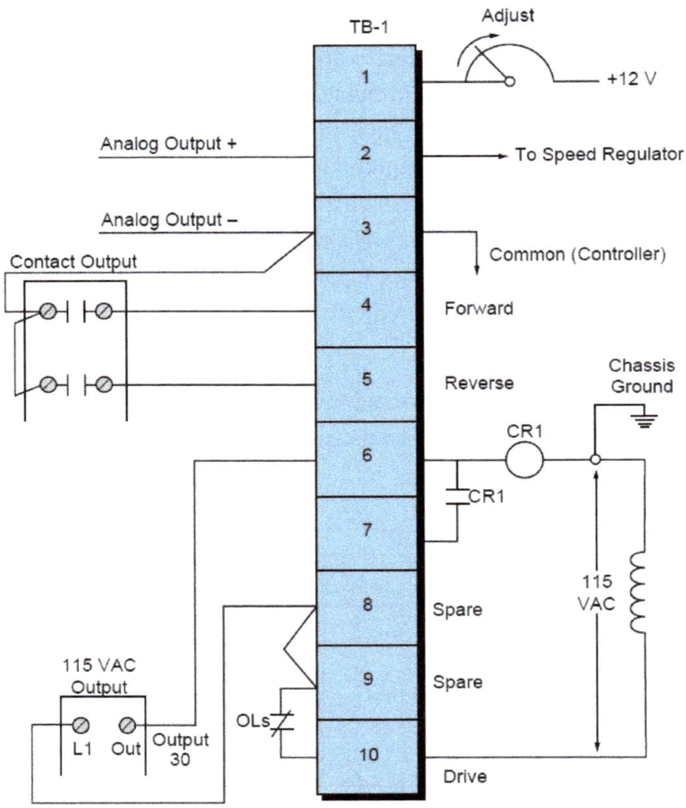

***Figure 40**: Connection diagram from the PLC to the VS drive's terminal block.*

Since the common must also be switched, the connection uses a contact input interface to switch forward/reverse signals. Terminal TB-1-6 must be connected to 115 VAC in order to activate the internal relay CR1. To turn the drive on, the terminal block TB-1-8 connects to the L1 connector of the PLC. Terminal TB-1-6 is connected to the output of module (CR1). The drive's 115VAC signal is used for controlling the motor speed. It is connected to terminal TB-1-6.

This configuration allows the motor's overload contacts to be wired to terminals TB-1-9 and TTB-1-10. These are the drive's power (L1) and output interface's connection L1. The drive will shut down if it experiences overload. This is because the drive's contact CR1 will not receive power from its output module. The configuration does not offer low-voltage protection as the motor and drive will restart immediately after overloads have cooled off. Low-voltage protection must be provided by the PLC using the auxiliary contact of the drive (CR1 terminal TB-1-7), so it seals the stop/start circuit.

Figure 41 illustrates the PLC ladder program, which will replace the manual operator station. Because the forward and reverse inputs can only be ON at any one time, they are interlocked (i.e. they are mutually exclusive).

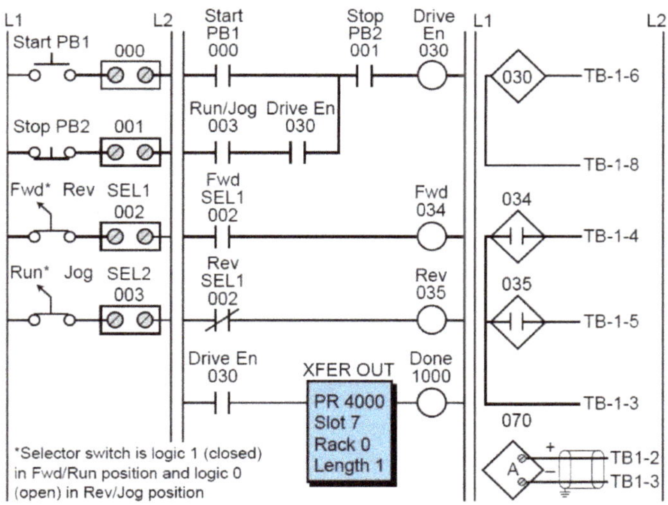

***Figure 41**: PLC implementation of the VS drive.*

The motor will run at the speed specified by the analog output when it is selected. The analog output connection allows for the output to be activated when the drive turns on. Register 4000 stores the count for the analog output to drive. Internal 1000 is used in block transfer to indicate completion of instruction.

Sometimes, VS drives require the ability to operate under manual or automatic control (AUTO/MAN). To implement dual control, several additional hardwired connections will be required. This can be done with a selector switching device (e.g., a single-throw, four-pole selector switch). It is the simplest and most cost-effective way to accomplish this. This switch allows the user to choose between an automatic or manual option. This connection is illustrated in Figure 42.

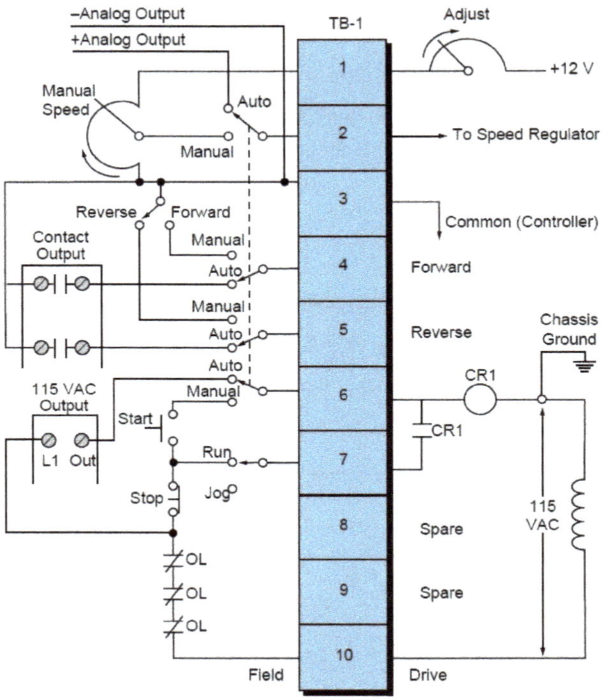

***Figure 42**: VS drive with AUTO/MAN capability.*

The operator station is where the forward/reverse field and start/stop buttons are located. These devices are connected under the same names as the control program. (Refer to Figure 41). The PLC will control the drive if the AUTO/MAN switch has been set to automatic. If it is set to manual, then the manual station will control it.

5.7. Continuous Bottle-Filling Control

This example (see Figure 43) will show how to implement a control program. It detects the position and waits for 0.5 seconds before filling the bottle with water until the photosensor detects that it is full.

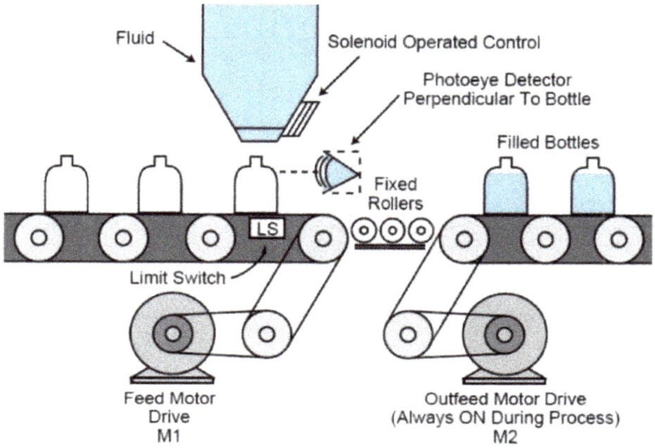

***Figure 43**: Bottle-filling system.*

The control program will wait for 0.7 seconds after the bottle has been filled before it moves on to the next bottle. This program will contain start and stop circuits to the outfeed motor as well as the start and end of the process. Table 17 lists the I/O address assignment. Tables 18-19 and 19 list the register and internal assignments. These assignments also include the start- and stop signal.

Module Type	I/O Address			Description
	Rack	Group	Terminal	
Input	0	0	0	Start process PB1
	0	0	1	Stop process PB2 (NC)
	0	0	2	Limit switch (position detect)
	0	0	3	Photoeye (level detect)
Output	0	3	0	Feed motor M1
	0	3	1	Outfeed motor M2 (system ON)
	0	3	2	Solenoid control
	0	3	3	—

Table 17. I/O address assignment.

Device	Internal	Description
Timer	1001	Timer for 0.5 sec delay after position detect
Timer	1002	Timer for 0.7 sec delay after level detect
—	1003	Bottle filled, timed out, feed motor M1

Table 18. Internal output assignment.

Register	Description
4000	Preset value 5, time base 0.1 sec (1001)
4001	Accumulated value for 1001
4002	Preset value 7, time base 0.1 sec (1002)
4003	Accumulated value for 1002

Table 19. Register assignment.

Figure 44 shows the PLC ladder implementation for the bottle-filling app. The outfeed motor (output 031) turns ON once the start button is pushed. It will remain ON until the stop button is pushed. Once the system starts (M2 ON), the feed motor M1 will turn ON until the stop button is pushed. It will stop once it detects that the bottle is in the correct position. After the bottle has been in place for 0.5 seconds, the solenoid 032 will open the filling vessel and stay ON until the photoeye(PE) detects the proper level. After the bottle is in place for 0.7 seconds the internal 1003 will turn on the feed motor. The feed motor will stay ON until another bottle is detected by the limit switch.

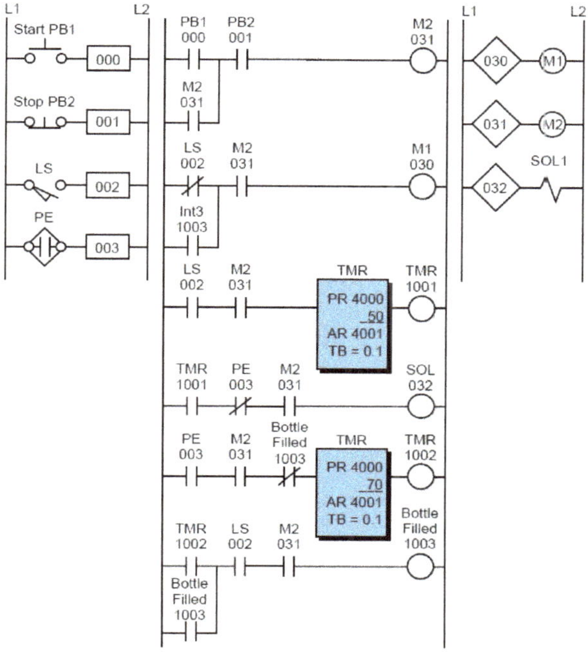

***Figure 44**: PLC implementation of the bottle-filling application.*

5.8. Large Relay System Modernization

This is an example of modernization for a machine control system. It will go from hardwired relay logic, to PLC programmed logic. With the exception of the ones that the controller can execute (e.g. timers and control relays), the field devices will not change. Modernizing this machine's control system will bring many benefits:

* A more reliable control system

* Less energy consumption

* Less space is required for the control panel

* A flexible system that can be expanded in the future

Figure 45 shows the relay ladder diagram which controls the logic sequence of this machine. The diagram only shows a portion of the relay ladder logic.

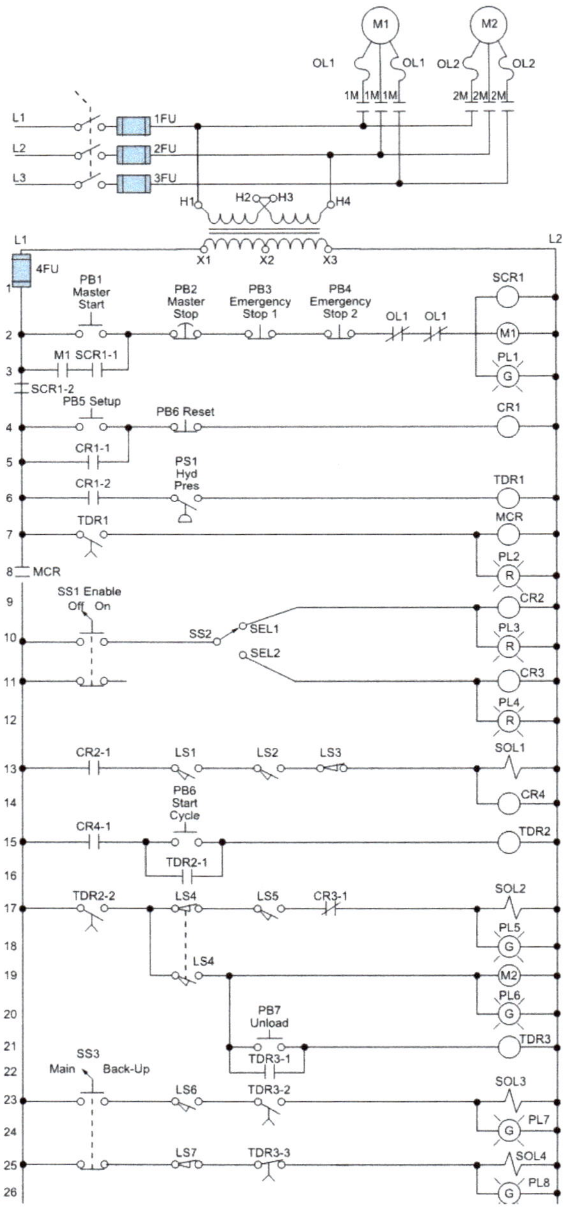

***Figure 45**: Electromechanical relay diagram.*

A quick review of the relay ladder diagram shows that some portions of the logic should remain hardwired. These are lines 1, 2 and 3. This will ensure that all emergency stop conditions are independent of the controller. Hardwiring the hydraulic pump motor (M1) should be done as well. It is only energized when master start push button (PB1) is pushed. These hardwired components are illustrated in Figure 46. The safety relay (SCR), which provides power to the rest, will only work if M1 is functioning properly and the emergency push button is not pressed.

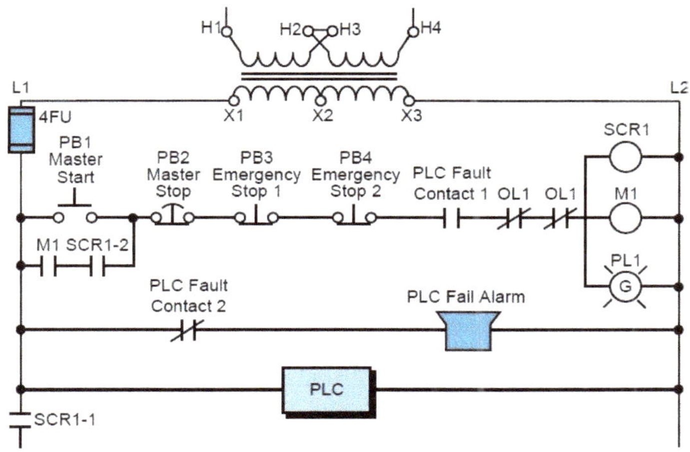

Figure 46. Elements of the modernization example system to be left hardwired.

The PLC fault contact can also be connected to an alarm for a PLC failure. The fault coil will be energized by the PLC during normal operation. This will close PLC Fault Contact 1 (closing) and open PLC Fault Contact 2.

Continuing with the example, let's now assign the actual inputs and outputs of the I/O assignment file. We will assign all control relays their internal output addresses, along with timers and interlocks. The assignment and description of inputs and outputs as well as internals are shown in Tables 20-21. You will notice that inputs with multiple contacts (e.g. LS4 or SS3) have one connection to the controller.

Module Type	I/O Address Rack	Group	Terminal	Description
Input	0	0	0	PB5—Setup PB
	0	0	1	PB6—Reset (wired NC)
	0	0	2	PS1—Hydraulic pressure switch
	0	0	3	SS1—Enable selector switch (NC contact left unconnected)
Input	0	0	4	SEL1—Select 1 position
	0	0	5	SEL2—Select 2 position
	0	0	6	LS1—Limit switch up (position 1)
	0	0	7	LS2—Limit switch up (position 2)
Input	0	1	0	LS3—Location set
	0	1	1	PB6—Start load cycle
	0	1	2	LS4—Trap (wired NC)
	0	1	3	LS5—Position switch
Input	0	1	4	PB7—Unload PB
	0	1	5	SS3—Main/backup (wired NO)
	0	1	6	LS6—Maximum length detect
	0	1	7	LS7—Minimum length backup
Output	0	3	0	PL2—Setup OK
	0	3	1	PL3—Select 1
	0	3	2	PL4—Select 2
	0	3	3	SOL1—Advance forward
Output	0	3	4	SOL2—Engage
	0	3	5	PL5—Engage ON
	0	3	6	M2—Run motor
	0	3	7	PL6—Motor run ON
Output	0	4	0	SOL3—Fast stop
	0	4	1	PL7—Fast stop ON
	0	4	2	SOL4—Unload with backup
	0	4	3	PL8—Backup ON

***Table 20**: I/O address assignment.*

Device	Internal	Description
CR1	1000	CR1 (Setup Rdy)
TDR1	2000	Timer preset 10 sec register 3000 (accumulated register 3001)
MCR	MCR1700	First MCR address
CR2	—	Same as PL3 address
CR3	—	Same as PL4 address
CR4	—	Same as SOL1
—	1001	To set up internal for instantaneous contact of TDR2
TDR2	2001	Timer preset 5 sec register 4002 (accumulated register 4003)
—	1002	To set up internal for instantaneous contact of TDR3
TDR3	2002	Timer preset 12 sec register 4004 (accumulated register 4005)

Table 21: Internal address assignment.

Figure 47 illustrates the PLC program code (hardwired relay translator) in this example. The ladder program illustrates the special coding techniques required to implement PLC logic. These techniques include the software MCR function and instantaneous contact from timers.

The program software specifies an MCR's internal output. This function is similar to a hardwired one. The relay logic diagram shown in Figure 45 shows that if the MCR has been energized, the contacts will close and allow power to flow to other parts of the system. The internal MCR 1700 in the PLC software performs the same function. In this example, MCR1700 represents the first address for MCRs. The PLC cannot execute the ladder logic between the MCR coil instruction and the ENDMCR instruction if the MCR coil has not been energized.

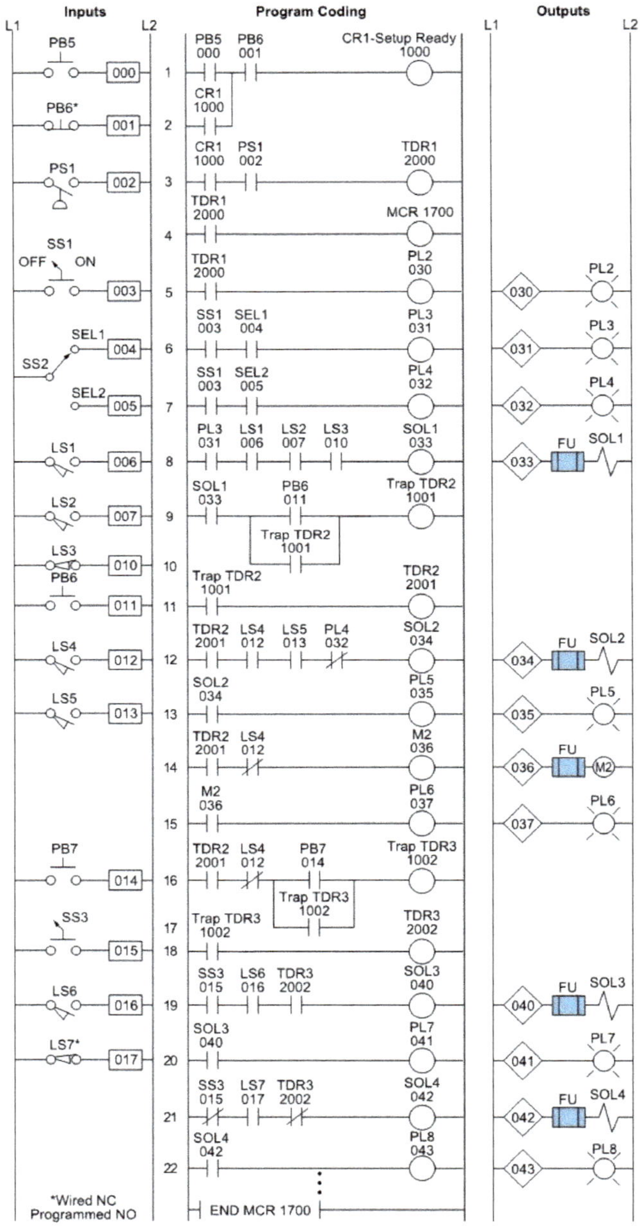

***Figure 47**: PLC implementation of the circuit in Figure 45.*

The control relay CR2 (line 9) cannot be replaced by an internal, as the PL3 contacts can be used in line 10. This can be used when a control relay is connected to a real output device. We don't need to separate the coils of lines 17 and 18. This is already done because the PLC here doesn't allow multiple output rungs. It is a good idea to use separate rungs for each output.

As explained in previous sections, the normally closed inputs connected to the input modules can be programmed as normally opened. Two contacts make up the limit switch LS4: a normally open and a normally shut one, respectively. These are shown in Figure 45 lines 17 and 19. Only one set of contacts must be connected to the controller. In this case, the normally closed contact LS4 has been selected. The normally open contact isn't connected to the controller but its hardwired function may still be achieved by programming LS4 in the form of a normally closed ladder touch.

These applications require timers that have instantaneous contacts. This feature is not available in all PLCs. A timer can be set to enable an instantaneous contact, which means that the contact opens or closes automatically when it is activated. An internal coil can be used to replace an instantaneous contact in most PLCs. Line 15 of the hardwired logic explains that if PB6 and CR4 are pressed, the timer TDR2 will begin timing and contact TDR2-1 seals PB6.

This arrangement requires special PLC implementation. Software timer contacts are required. The timer will not seal until the timer has run out. The timer will reset if PB6 gets released. This is because PB6 has not been sealed. This problem can be solved by using internal coil 1001 for sealing PB6. Then, the TDR2 timing timer will start. This technique is shown on lines 9, 10 and 11 of PLC program coding. ON delays are handled by the time delay contacts (2001).

6. Industrial Applications of PLCs

PLCs are programmable logic controllers that are designed for fast cycle times and interaction with the external environment. An Allen-Bradley PLC system includes a left computer module and two additional input and output modules. Its purpose is to replace complicated and slow wiring systems that require electricians. This need led to the creation of ladder logic language. Now, PLC programming can be done by any individual with a little experience in programming.

6.1. Process Control

With the introduction of intelligent automation, robotics and the Internet of Things, the manufacturing process has become much more efficient. While PLCs have always played a vital role in the manufacturing process, new technologies are making them more efficient than ever. For instance, the power of cloud computing is enabling these devices to synchronize with high-level computing systems, making the entire production process more efficient. The data collected from these devices can also be analyzed to produce a full-blown picture. The resulting analysis can help plant managers better leverage resources and develop efficient manufacturing processes.

The PLC is an electronic device that monitors the inputs of specific devices. It measures the data against pre-defined parameters and rules and decides whether or not outputs should be triggered. The PLC's CPU serves as the brain of the machine, evaluating data according to pre-defined rules to generate actions that control output devices. The data is then communicated throughout the organization through a network of communication equipment. An HMI interface allows technicians and managers to access information from the PLC system.

The PLC has built-in ports that enable communication with external systems and devices. This communication is carried over various industrial network protocols, such as EtherNet/IP or Modbus. For larger I/O systems, peer-to-peer communication is also possible. This communication enables separate parts of a complex process to be controlled independently. The communication links are often used for HMI devices. If a communication link breaks down, a PLC will automatically switch to another communication channel to ensure the process continues to run as planned.

The PLC is an important piece of equipment for industrial manufacturing. Not only can it improve productivity, but it can also improve inventory management. If inventory is not organized properly, inventory can get misplaced and lost labor costs will increase. These benefits are obvious, but what makes PLC systems so powerful is the versatility they offer. Oil and gas companies often add well pad sites to their expansion plans. This gives them more opportunities to drill more than one well, and increases the amount of production per area.

6.2. Safety-Critical Applications

When you think of safety-critical applications, what comes to mind? Perhaps you think of manufacturing plants or other industrial facilities, which need to run without interruption. In these cases, PLC programming can be vital to keeping the facility running as smoothly as possible. This article explores safety-critical PLC programming and its importance to the industry. Read on to learn more. Safety-critical applications require PLC programming in industrial settings.

Using symbolic execution and dynamic simulations to detect PLC safety violations can help ensure that systems are safe. Real-world PLCs never work by themselves, but rather collaborate with other programmable components. As such, the logic of the PLC is driven by external events that can affect the safety of the system. This type of programming can lead to safety violations and must be protected against. To combat this problem, PLC programming must be dynamic and temporally aware.

The challenge of analyzing PLC code is not unique to PLCs, but rather specific to them. This is because it controls several types of customized hardware and processes. This software follows a unique programming paradigm and introduces a number of new concepts, including PLC scan cycles. This sensitivity makes PLC events extremely time-sensitive. Because of this, the safety of a PLC program is dependent on its reliability.

A safety PLC must be able to detect 99% of potential failures. This is done through diagnostic coverage tests, such as data verification and program flow control. Both of these tests verify that the PLC's internal functions are in order. Safety PLCs must also undergo software-fault injection tests, which download a corrupted program into the PLC. If these programs fail, the PLC must respond safely.

A safety PLC may need additional redundancy to be effective. For example, the Siemens S7-400/300 series utilizes dual-processors, which ensures safety reliability by ensuring that the safety PLC is programmed with at least two microprocessors. In addition, the safety PLC has dual processors that execute in parallel, one in 16-bit word mode and the other in single-bit binary instruction mode.

6.3. Remote Monitoring

For remote monitoring and control of PLCs, Ursalink provides a cellular router and an intelligent PLC management solution. The UR72 industrial cellular router helps reduce part cycle time, improve quality, and decrease production cost. This remote monitoring and control software is easy to install on the computer of your PLC operator. To learn more about how it works, please read on. This article will provide you with an overview of how the technology works.

For remote monitoring of PLCs, you can use several protocols. These protocols include MORDEN, OPC, and ETHERNET. In addition to these protocols, some remote monitoring software includes a web portal.

M2Web can be accessed from mobile devices, as well. This software is also mobile-specific, so you can use it on the go. In this article, we will discuss how these technologies work to enable remote monitoring.

The first major limitation of remote monitoring is that there is no direct access to the core data. For this reason, it is necessary to plan the number of web servers and hardware/software data points to ensure the best remote monitoring system. Furthermore, the remote systems need additional web servers and peripheral resources, which will further complicate the structure of the remote monitoring system. Nevertheless, the architecture proposed in this article is web-based and all functions can be realized directly on the web.

M2Web is a cloud service and hardware bridge that enables manufacturers to remotely access and program PLCs. It can also read different types of data from various PLCs. The service also has the capability to store this data for notifications. To use M2Web, you must first create an account with Talk2M, a free cloud service that can create powerful industrial connectivity. Then, you will be able to monitor, manage, and configure the PLC remotely.

Internet TRiLOGI works like any Windows-based program. It is based on client/server architecture. The internet connection enables you to access the PLC from anywhere using a Java web browser. You don't need to install any local software to connect to the PLC via the Internet. You can even connect the PC to the modem using standard TBASIC commands. It is that easy. The TLServer 2.0 software suite can be used to remotely monitor PLCs.

There are more industrial applications of Programmable logic controllers. These controllers are ideal for a number of reasons, including reliability, simplicity, and low maintenance costs. For instance, Cement industry PLCs can control access to a robot cell with trapped-key access. SCADA systems can also be used to monitor and control industrial processes. In addition, Modular PLCs can be installed in an industrial setting.

6.4. The Cement industry

Cement industry uses the PLCs. The manufacturing of cement involves the use of PLC systems to mix raw materials inside a kiln. The proportions of these raw materials play an important role in the final product's quality. The cement industry utilizes PLC in configuration software and user mode for the process. PLCs also operate in ball milling, coal kiln, and shaft kilns. This article will discuss some of the common PLC applications in the cement industry.

The first application of PLCs is in the cement kiln. The PLC programme allows users to monitor other processes in real-time. This function is particularly useful for monitoring the operation of multiple processes simultaneously. Moreover, it allows users to change the maximum operating temperature of each motor. The Cement Kiln PLC system has a graphic window to display its operating temperature. The operator can change this temperature manually or through SCADA.

Another area of application for PLCs is in the paper industry. For example, they can be used to control high-speed paper machines and monitor the production of pages. Cement is a complex chemical product that is produced by combining several raw materials in a kiln. The proportions of these raw materials play a key role in the final product. Accurate data on these variables is vital in determining the proper raw materials for a particular process.

6.5. Safety PLCs Control access to a Robot Cell with Trapped-key Access

A safety PLC can be used to control the access to a robot cell with trapped-access keys. This type of control can also manage shutdown responses in case of an emergency stop. A safety PLC is typically a standalone unit with restricted regular instructions, but it can also be added to existing controller architectures for safety-critical applications. They are ideal for small, lightweight machines and are capable of executing PLC languages, such as Ladder and Flow-Chart/Grafcet. Safety PLCs are easy to maintain and integrate into custom printed circuit boards.

An example of a trapped-key access PLC is one which allows operators to enter and leave the hazardous area by pressing a button. The safety interlock switch will trigger a process which will prevent the robot from moving until the operator has entered the cell and released the trapped-key access. When the door opens, the safety relay trips. This means that the operator cannot enter the cell and cause a safety hazard.

Trapped-key interlocking is another type of mechanical lock that restricts access to certain areas. Trapped-key access relies on the practical principle that "a key can't be in two places at the same time." In simple terms, the trapping mechanism allows a person to enter the dangerous area only when the appropriate key is inserted. This transfer of keys facilitates safe sequences of operations and protects the process and the product in production.

6.6. SCADA Systems Control and Monitor Industrial Processes

Industrial processes, such as refining, manufacturing, or power generation, are often monitored and controlled using SCADA systems. These processes may run in a continuous, batch, repetitive, or discrete manner, depending on the system. They also include public and private processes, such as water and wastewater collection, oil pipelines, and wind farms. SCADA systems can control and monitor a range of factors, from HVAC systems to water levels at dams.

In addition to controlling and monitoring industrial processes, SCADA systems allow users to monitor and analyze real-time data from a wide range of sources. This allows machine operators and technicians to be in one location while still allowing them to perform limited tasks. In addition to reducing human error and labor costs, these systems also increase productivity and efficiency, extending equipment life, and ensuring regulatory compliance. These systems allow organizations to control and monitor their processes in real time and store data to help them improve their operations.

SCADA systems operate at different levels. The first level, called Level 0, comprises sensors and industrial equipment that provide data to a controller. Level 2 is the functional level, which provides the supervisory computer with visualization for dedicated HMIs. At this level, data collected by the controllers are fed into a database server for analysis in the future. Level 3 is the production control level, which does not directly control the field devices, but instead concentrates on production monitoring. HMIs, located within a plant, can monitor production.

6.7. Modular PLCs

PLCs are programmable control systems, which apply ladder logic to various inputs and outputs. Their outputs can be connected to mechanical, hydraulic, or manufacturing equipment. They can also communicate with one another through a variety of digital communications methods, such as PROFIBUS, PROFINET, or Modbus. Listed below are some benefits of using PLCs in industrial applications. Read on to learn more.

Modicon developed the first Programmable Logic Controllers (PLCs). They reduced cabinet space and eliminated the need for rewiring. They also improved functionality and lowered installation costs. Michael Greenberg designed the first commercially successful PLC, the Modicon 184, in 1973. Since then, the technology has been widely used in many industries. To date, Modicon has sold over five billion units.

Modern PLCs have improved their memory and processor speeds. Many have become more efficient than their predecessors. Megabytes are now the new standard for measuring memory, and the introduction of User-defined data types allowed for the manipulation of machine data and the sharing of information. These controllers are easy to learn and are ideal for beginners. In addition to improving functionality, today's PLCs take advantage of rapidly declining solid-state memory to increase speed and memory capacity.

The benefits of PLCs extend beyond their use in industrial settings. In aerospace, they can control water tank quenching systems, food-process filling machines, and closed loop textile shrinkage control systems. In the food industry, they can control silo feeding and injection molding. Their use is increasing and standardizing processes, improving returns on investment. The following are some benefits of using PLCs in industrial applications.

6.8. Benefits of Using PLCs in Industrial Applications

Compact PLCs are Cheaper than a PC

While both PLC and IPC are cheaper than a PC, PLCs have several advantages over their PC counterparts. Both require I/O modules and HMI software, but PLCs can also deploy a GUI directly on the controller. The other main difference between the two is their lifecycle. Generally, PLCs have longer support lifecycles, but they are also more expensive to develop. PLCs can achieve elegant solutions by implementing ladder-logic programs.

The cost difference is largely due to the difference in programming languages. The graphical language is used more frequently in industrial applications, allowing for more complex functions and diverse I/O ports. It is also easier to learn than the textual language, so it is the preferred choice for most PLC projects. Beginners can get up to speed with the graphical language within 40 hours of training, or even a single week.

PLCs can be extremely versatile and cheap. They can automate interior and exterior lights, train layouts, and many other simple tasks. A PLC can also be an invaluable learning tool for ladder logic and automation. If you are new to the field of industrial automation, consider a PLC to learn the basics of automation. They can also be used in manufacturing to automate train layouts and train interiors.

Communication with a PLC

As you must have known, a PLC is an electronic device used to control various machinery. They work by processing inputs and executing program instructions to deliver results. They typically function in a loop, processing inputs, executing program instructions, and then delivering results. The structure of a PLC is very similar to that of a computer. This section will cover some of the basic aspects of communication with a PLC.

There are three main methods for communicating with a PLC: ladder logic, Boolean, and Grafcet. The first two methods are the most commonly used, and are both common. The second method is more complex, utilizing functional blocks that carry out operations when enabled by an input. The latter approach does not allow for the creation of sneak paths. A ladder diagram will show the various methods of communication.

In addition to communication with a PLC, the controller must have enough memory to store its data. Most PLCs are designed for safe operation at temperatures between 0 and 60 degrees Celsius. Specialized models are available for extremely cold or hot temperatures. The PLC should have sufficient RAM and ROM to store and execute its operating system. It should have enough input and output ports to connect with peripherals.

A ladder-diagram program is stored in the processor. Once stored, it is in the processor until changed. If the power source goes off, the ladder-diagram program remains unchanged. Therefore, it is critical to understand the basics of ladder-diagram programming before starting a project with a PLC. Communication with a PLC can also be done through communication networks. Communication with a PLC for industrial applications of Programmable logic controllers.

6.9. Application of PLCs to Paper Mill, Coal Kiln, Shaft Kiln and Glass Industries

You might be wondering, how do PLCs help in the Paper Industry? Well, in this section I will provide you with information on how PLCs help in paper mills, coal kilns, and shaft kilns. Read on to find out!

The Paper industry makes use of a lot of PLCs. But how do they work? Here is an overview of some common PLC applications in the Paper Industry.

6.9.1. How PLCs are Used in Ball Mill

The process of charging the ball mill is based on a control system that can regulate the flow of the mill's energy. The drive system is called the ball mill drive system, and it does not have a gearbox. The ball mill is charged with an electric current using the SAS motor, which is operated at six kV. The power of the motors is determined by their voltage, and the range is from 400 to 1120 kW. The nominal efficiency of the mill is around 80 Mg/h.

For instance, in the paper industry, a PLC controls the material ratio, the flow rate, and the temperature in a mill. This machine also processes flat glass. The advanced technology of the PLC makes it highly desirable in many fields. In water treatment plants, a PLC controls the flow rate and contaminant levels in major parts of the plant. The PLC also controls the valves controlling the flow of in-process water. It also controls the creation of newspaper and book pages.

As you must have known by now, the PLC consists of basic elements, such as a CPU, input modules, and programming devices. Input modules respond to analog and digital signals. The input circuitry converts these signals to logic signals. The CPU, or central processing unit, evaluates the state of the inputs and outputs by evaluating them.

These factors affect the process's efficiency. When a PLC fails to respond to a request, a machine's outputs and variables will fail to meet production goals.

6.9.2. Coal Kiln

In the coal kiln industry, PLCs are a vital part of the overall process automation. These intelligent machines help to optimize various aspects of a coal kiln's firing and control. For example, a coal kiln's air flow depends on various process variables. A PLC in this industry controls the air flow, which varies depending on the material's chemical composition and moisture content.

Direct-fired rotary kilns use refractory to protect materials from direct contact with combustion products. An indirect-fired rotary kiln, on the other hand, utilizes a heating shroud to prevent the material from coming into contact with the combustion products. The heating medium is introduced into the kiln's drum, which constantly rotates, redistributing the bed of material. Added internals can optimize processing.

In the **cement industry**, a coal kiln is used to mix several raw materials and the proportions of each are important to the quality of the finished product. Using a PLC in the coal kiln industry can help manufacturers ensure that the process is efficient and profitable. This technology is used in many industries, including cement, shaft kiln, and ball milling. In addition to optimizing the operation, PLCs help maintain the desired temperature and fuel/air ratio, thereby reducing the risk of a fire hazard.

The most important parameters that a PLC in the coal kiln controls include clinker production rate, fuel flow rate, specific heat consumption, secondary air temperature, and preheater exhaust gas temperature. In addition, the PLC program continuously monitors the preset values of the inputs and updates the outputs. The ID fan speed controls stirring and heat exchange in the kiln.

6.9.3. Shaft Kiln

A distributed control system consists of a PLC in user mode and configuration software that manages all processes in the cement, coal, and paper industries. It is used in kilns and coal mills to control the process parameters, including the lighting of the main burner, fuel changes, and temperature. These systems are also used in civil and aerospace applications, such as traffic signals. They also control water tank quenching systems and ball mills.

A PLC is a great choice for paper mills for a number of reasons. Its cost is far lower than custom software, has a proven track record, and is backed by a solid support team. PLCs are also very stable, meaning they can last for decades. They do not require any special training or maintenance, which means employees will not have to perform extensive work to monitor and maintain the system.

PLCs can be programmed in a variety of different languages, including graphical and text-based. Those that have a higher degree of experience using such technologies are the best candidates. Graphing-based languages are preferred by many clients. Alternatively, PLCs for paper mills can be written in a text-based language. In any case, the programming language must be able to support a number of different languages.

6.9.4. Paper Mill

Paper mills use programmable logic controllers (PLCs) for a variety of processes, including receiving raw material, processing it, and storing it. Often, large equipment is involved, and a PLC can control the system at any level to control the various processes. PLCs can be configured with functions such as start/stop, count sequential, interlock, and time delay. They can also be multiplexed, which allows multiple signals to be sent over the same pair of wires.

Earlier, PLCs could not be programmed, but modern PLCs can be programmed by using a computer. This automation allows paper mills to control numerous processes at once, resulting in greater efficiency.

The paper mill industry uses PLCs to control the flow of in-process water, monitor water contaminant levels, and control other processes. These sophisticated machines can also monitor water filtration processes and control valves that control the in-process flow of water.

A PLC offers advantages in many aspects of paper mill automation, including a reduced cost and more robust computing system. PLCs take up a much smaller footprint than the functional equivalent, so PLC units can be used virtually anywhere. In addition, PLCs are safer and more secure than desktop computers, which can be prone to damage. These factors, combined with the fact that PLCs are easy to program, make them the ideal solution for paper mill automation.

Safety PLCs

As the number of inputs in a paper production line increases, safety-control systems are becoming more complex and need to monitor more variables. Safety PLCs provide a unique advantage over safety relays in this respect: they combine standard functionality in a single controller. These controllers enable manufacturers to reduce the complexity of the overall control system, as well as its size and design time. The following are some of the benefits of safety PLCs in the paper industry.

6.9.5. Glass Industry

Glass production is an elaborate process. Glass companies typically use PLCs with bus technology. They use PLCs in digital quality control, position control, and machine monitoring. Safety PLCs are also used in the paper business for a variety of processes, including controlling machines that are moving at high speeds and monitoring the production process. Ultimately, safety is a priority for this industry, so it is crucial to utilize these technologies wisely.

Another benefit of safety PLCs is that they can provide advanced diagnostics. For example, safety PLCs perform data verification and program flow control to ensure that essential functions execute in the right order. In addition, data verification stores critical data redundantly in memory and verifies its validity before use. These features require additional software testing techniques, such as software-fault injection tests, to ensure that all of the required safety functions are fully implemented.

6.9.6. Benefits of PACs and PLCs in Industries

PLCs, or programmable logic controllers, are used in many industries, from the glass industry to the paper industry. In both industries, they automate and monitor many processes. In the glass industry, PLCs are used for digital quality control, positioning control, and many other tasks. In the paper industry, PLCs are used in different processes, including controlling machines at high speeds, monitoring product on, and ensuring efficient processes.

The benefits of PACs outweigh any potential drawbacks. The PACs are often redundant and use open industry standards like PROFINET tc reduce downtime and improve performance. These controllers also feature fanless designs to reduce the risk of failure and increase machine uptime. These advantages make them an excellent choice for paper and pulp manufacturing operations. Emerson's programmable automation controllers for paper and pulp production offer numerous benefits for paper and pulp companies.

The paper industry has a large number of challenges, so PACs are a good choice for many companies. These machines are designed to run continuously, reducing manual intervention and improving productivity. The PACs are often more affordable than PLCs, and their modular design makes them easy to add or remove functions as needed. In addition to being highly customizable, PACs are compatible with many enterprise networks. Moreover, PACs are ideal for industries with multiple processes and a variety of needs.

6.10. Safety PLCs vs Standard PLCs

A safety PLC is very similar to a standard PLC. It can be used to automate and control pieces of industrial equipment. A safety PLC supports all the applications that a standard PLC does. In addition, a safety PLC contains integrated safety functions that allow it to control safety systems as well.

6.11. Safety Precautions to Take When Working with PLCs

There are a few basic safety precautions to follow when working with PLCs. The environmental conditions that surround these components are essential for the proper operation of the components. To avoid downtime and minimize the risk of foreign signal interference, the environment should be as clean as possible. Dust on the circuit boards could lead to catastrophic short circuits. Keep the enclosure filter clean to maximize airflow and ensure consistency. When working with PLCs, it is critical to follow all manufacturer's safety guidelines.

PLCs are designed to enable safety functions to be realized

In addition to its reliability, a safety PLC meets rigorous international standards. IEC 615081, for example, addresses functional safety of electrical and programmable electronic safety-related systems. It covers design methods and hardware tests and acknowledges that zero-risk is impossible to achieve. Third-party experts, including UL and TUV Rheinland, enforce the standards. To ensure that a safety PLC meets strict safety requirements, it should be installed in critical applications.

Moreover, safety PLCs are designed with special diagnostics for the implementation of safety functions. The diagnostics include data integrity checking and program flow control. Program flow checking ensures that essential functions execute in the correct sequence, while data verification stores critical data in redundant memory and verifies its validity before use. Furthermore, safety PLCs require extra software testing techniques such as software-fault injection tests. This technique deliberately corrupts the software to determine if the safety PLC can perform its required functions.

Safety PLCs are also known as programmable logic controllers (PLCs). They are special-purpose machines that offer critical control and safety applications. Safety PLCs typically form an integral part of safety instrumented systems that monitor processes and automatically bring them into a safe state. Although conventional PLCs have been used for years, they do not include the special features that are essential for safety functions.

Safety PLCs can be used instead of standard PLCs

Various safety measures need to be taken when working with PLCs. Safety PLCs are used in industries involving oil and gas, petrochemicals, refineries, power plants, incineration, boiler controls, and machinery. They are used in combination with safety I/O blocks, which handle the necessary safety inputs. Safety PLCs may be used instead of standard PLCs in some applications, but the benefits of using them are significant.

Safety PLCs must pass stringent international standards. The highest SIL level is three, meaning they must detect over 99% of all system failures. They are also equipped with diagnostics to prevent system malfunctions. While standard PLCs are designed for industrial use, safety PLCs are safer and more robust. They feature redundant microprocessors and are often used instead of standard PLCs.

Safety PLCs are more expensive than standard PLCs but can save money in the long run. However, they tend to have higher startup costs and are not suitable for small applications. Safety PLCs are also more complicated to install and wire than standard PLCs, so engineers and maintenance personnel may need additional training to ensure proper installation and operation. For these reasons, safety PLCs are the preferred choice for many applications.

When working with safety PLCs, it is imperative to adhere to the strictest international safety standards. Safety PLCs should meet IEC 61508, Functional Safety of Electrical/Electronic/Programmable Safety Related Systems. These standards provide the necessary diagnostic coverage for safety systems and minimize non-tolerable risks. In addition to this, safety PLCs also must meet standards that enforce strict certification procedures.

They have redundancy

There are numerous safety precautions to follow when working with PLCs. In addition to following the proper operating procedures, PLCs should be designed to fail safely and use silicon from separate lots. If possible, choose safety PLCs that are third-party certified to the EIC 61508 standard. Also, ensure that the interface of the safety PLC matches that of other PLCs. This will help minimize confusion and errors.

The environment around PLCs is often hazardous. Regularly checking the power supply is critical to the safe operation of the PLC. Additionally, make sure that the equipment is properly grounded. Electrical white noise barriers can help prevent distractions during faults. Loose connections are also common in electrical environments, so be sure to keep the area clean. Make sure that all wiring is secure. If you're working in a noisy or crowded environment, keep these safety precautions in mind.

Another important safety precaution to follow when working with PLCs is to avoid electromagnetic and radio frequency interference. The PLC can malfunction due to many factors, including a power outage. If this happens, the entire PLC system can be shut down. To fix the problem, the engineer must isolate the malfunctioning device and trace the problem back to it. Power outages, power fluctuation, or internal error can all cause problems.

They are sensitive to foreign signal interference

While working with PLCs, it is crucial to be aware of the possibility of foreign signal interference. In addition to the physical environment, other factors can cause interference as well, such as handheld radios, large motors, and arc welding. If the PLC is exposed to these types of signals, it can suffer damage that could be costly or even cause a downtime problem.

In addition to the physical environment, electrical noise can cause intermittent faults, weird behavior, and PLC failure. To prevent these issues, engineers should take necessary precautions before beginning work. These safety measures include conducting periodic inspections and testing the grounding integrity of the PLC. In the event that it is not completely intact, the engineer should test it with a multimeter to verify that it is grounded correctly.

Another safety measure is to ensure the quality of the power supply system for the PLC. It must be powerful enough to accommodate the load. If it isn't, the engineer can trace the source of the problem. In addition, PLC configuration errors can cause electromagnetic and radio frequency interference. Therefore, the engineer should follow the issue back to the source. The engineers should also ensure that the PLC is installed with sufficient emergency circuits, which can shut down the system manually if necessary.

Generally, PLCs are designed to control a series of devices and processes. In addition to monitoring individual machines, they can control a complex electro-mechanical process. With their multiple outputs, they can be linked to multiple parts of a system and program it to send an output signal to another. However, if there are foreign signals present, the safety precautions to take while working with PLCs are extremely important.

They are flexible

There are many safety precautions to consider when working with PLCs. First, make sure that the area is clean and dust-free. Dust accumulates on industrial settings and can damage the hardware inside the PLC. To reduce dust, consider installing a low-dust enclosure with adequate ventilation. Cleaning the enclosure regularly will prevent dust buildup. Keep all electrical components clean. If possible, try to avoid using compressed air.

If you plan to use handheld transmitters or antennas in close proximity to PLC equipment, make sure to isolate those devices. EMI can cause expensive damage and can result in significant downtime. Grounding is another important safety precaution. Electrical white noise barriers can be installed around machinery to prevent distraction during faults. Keep in mind that many electrical environments are susceptible to electromagnetic interference. Large motors, handheld radio systems, and radio antennas are all potential sources of electromagnetic interference.

Safety PLCs must comply with international standards for both design and functionality. The highest level of safety integrity is SIL 3, which requires the PLC to detect 99% of potential failures. They also undergo many diagnostic tests. The safety features on these PLCs are much more sophisticated. If you are in the oil and gas industry, be sure to invest in a safety PLC. These devices are available in a variety of brands, from Siemens, Omron, Beckhoff, Mitsubishi, and Schneider.

They are easy to troubleshoot

There are some safety precautions to be aware of when working with PLCs. You should know that red color fault LEDs mean that the CPU module is malfunctioning. To troubleshoot a PLC, you need to know its fault description and symptoms. Then, divide the troubleshooting procedure into sub-sections. If the fault has to do with manual mode, test the operation manually. Manual mode operation of a PLC is simpler than that of an automatic system.

When troubleshooting a PLC, you should follow a preventative maintenance plan to ensure that it is in good working order. This preventive maintenance plan should include system updates and considerations such as the size of your maintenance staff. Basic maintenance tasks should be performed daily. These include cleaning, tidying, and backing up data. It is advisable to consult an expert for complex troubleshooting procedures.

A PLC's memory can be corrupted by external factors. Such factors can include power interruptions and frequency interference. Corrupted memory can prevent it from executing a program correctly. It is recommended to back up all program data to a backup storage device and store it away from high humidity and RFI EMI exposure. If this fails, reinstalling the program is the best way to correct the issue.

7. Study Guide & Review Questions

The first step in developing a control program is the definition of the *control task*. The control task specifies what needs to be done and is defined by those who are involved in the operation of the machine or process.

• The second step in control program development is to determine a *control strategy*, the sequence of processing steps that must occur within a program to produce the desired output control. This is also known as the development of an algorithm.

• A set of guidelines should be followed during program organization and implementation in order to develop an organized system. Approach guidelines apply to two major types of projects: new applications and modernizations of existing equipment.

• *Flowcharting* can be used to plan a program after a written description has been developed. A flowchart is a pictorial representation of the process that records, analyzes, and communicates information, as well as defines the sequence of the process.

• Logic gates or contact symbology are used to implement the logic sequences in a control program. Inputs and outputs marked with an "X" on a logic gate diagram represent real I/O.

• Three important documents that provide information about the arrangement of the PLC system are the I/O assignment table, the internal address assignment table, and the register address assignment table.

- The *I/O assignment table* documents the names, locations, and descriptions of the real inputs and outputs.
- The *internal address assignment table* records the locations and descriptions of internal outputs, registers, timers, counters, and MCRs.
- The *register address assignment table* lists all of the available PLC registers.

• Certain parts of the system should be left hardwired for safety reasons. Elements such as emergency stops and master start push buttons should be left hardwired so that the system can be disabled without PLC intervention.

• Special cases of input device programming include the program translation of normally closed input devices, fenced MCR circuits, circuits that allow bidirectional power flow, instantaneous timer contacts, and complicated logic rungs.

- The programming of contacts as normally open or normally closed depends on how they are required to operate in the logic program. In most cases, if a normally closed input device is required to act as a normally closed input, its reference address is programmed as normally open.

7.1. Problems & Solutions to Build Your PLC Skills

7.1.1. Review, Reinforce, Test & Sharpen your PLC Skills

Solve the following problems to review, reinforce, test and sharpen your PLC skills.

1. What is the first step in designing an effective PLC control system?

A. approach the system in a systematic manner
B. flowchart the process
C. define the control task
D. define the control strategy

2. A(n) ______ is a procedure that uses a finite number of steps to achieve a desired outcome.

3 List four guidelines that are recommended as an approach to modernizing a control system.

4. In a modernization project, an existing ____ often defines the sequence of events in the control program.

5. System operation for new applications usually begins with:

A. sample diagrams
B. specifications
C. the control strategy
D. logic diagrams

6. A(n) ____ is a graphical representation of a solution's algorithm.

7. Logic sequences for a control program can be created using:

A. logic gates
B. relay ladder symbology
C. PLC contact symbology
D. all of the above

8. Draw the equivalent logic gate diagram for the circuit shown in Figure 1.

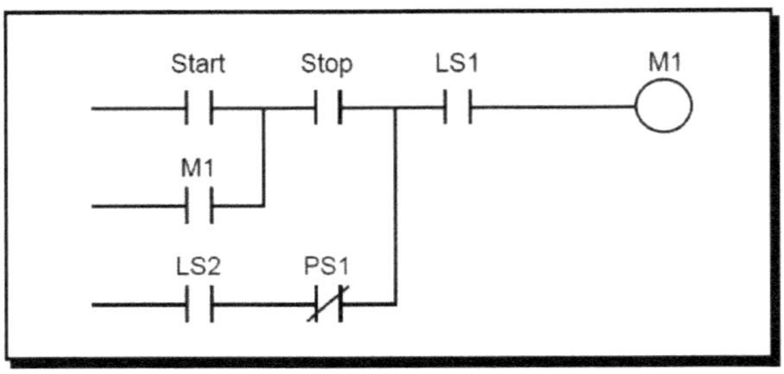

***Figure 1**. Circuit for problem 8.*

9. Draw the equivalent contact symbology diagram for the logic gates shown in Figure 2.

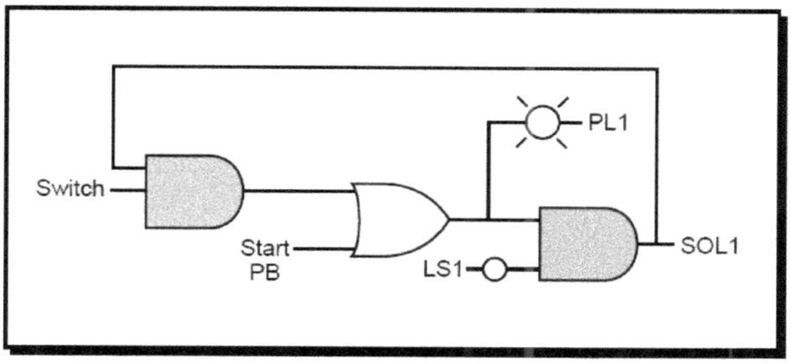

__Figure 2__. Logic gates for problem 9.

10. True/False. Only real inputs and outputs are documented during address assignment.

11. I/O address assignments are typically represented in one of three number

systems: _____ , ________ , or _______ .

12. The I/O address assignment table should closely follow the

_______ .

13. Using the circuit shown in Figure 3 and assuming that the PLC has a modularity of 8 points per module, there are eight modules per rack, the master rack is numbered 0, and the number system is octal:

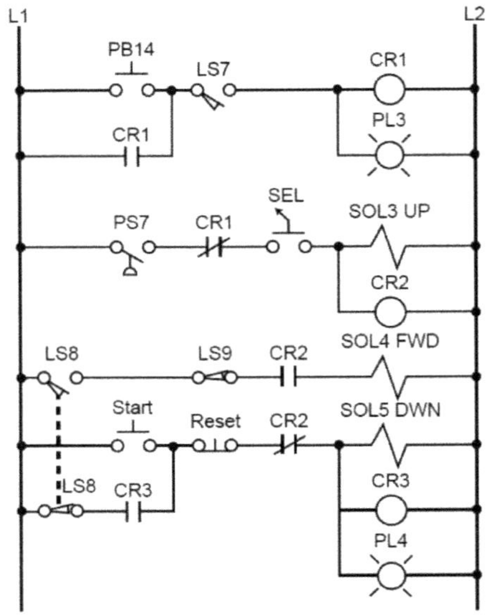

Figure 3. Circuit for problem 13.

(a). circle all real inputs and outputs

(b). assign the I/O addresses

(c.) draw the I/O connection diagram

14. The principal reason for leaving certain portions of the control circuit hardwired is to:

A. minimize wiring
B. avoid failure of main magnetic elements
C. ensure safety
D. keep some devices running at all times

15. The PLC fault contacts are wired to other hardwired emergency circuit elements:

A. in parallel
B. in series
C. normally open
D. normally closed

16. The main reason the PLC fault contacts are included in the hardwired circuit is:

A. to prevent system shut down
B. to detect I/O failures
C. to include the PLC as an emergency stop condition
D. to shut down the system if there is a PLC failure

17. Describe the purpose and operation of a safety control relay (SCR).

18. True/False. Normally closed input devices are always programmed normally open.

19. What is the purpose of the normally closed PLC fault contacts in the circuit in Figure 4 and describe what will happen if the PLC fails?

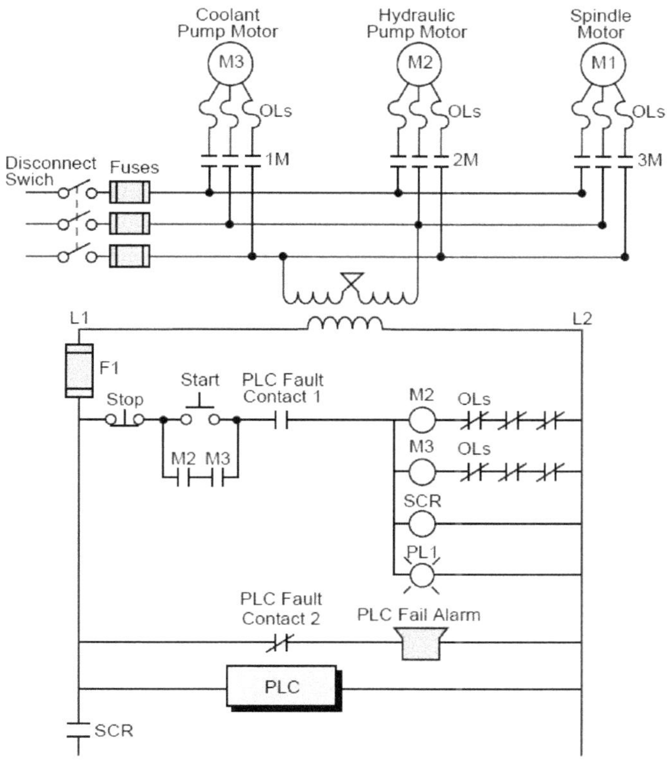

Figure 4: Circuit for problem 19.

20. Using the circuit shown in Figure 5 and starting inputs at address 10_8, outputs at address 50_8, and internals at address 100_8:

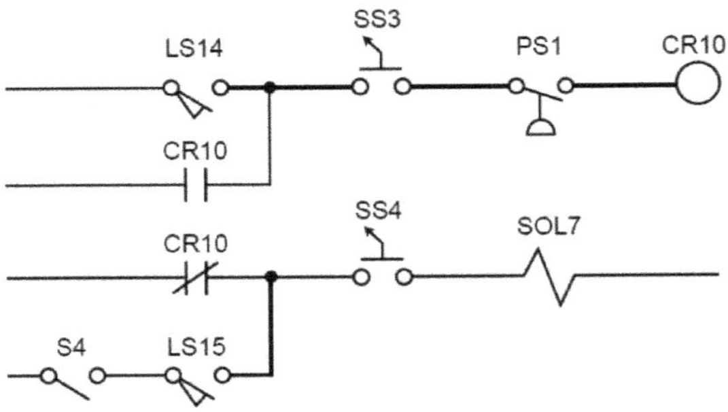

Figure 5. Circuit for problem 20.

(a). assign the I/O addresses

(b). draw the equivalent PLC ladder diagram

21. True or False. In a PLC ladder program, an END MCR instruction must be used to fence the area controlled by a master control relay.

22. What element can be used to trap timers in a PLC control program?

A. a reset instruction
B. a start push button
C. a pilot light
D. an internal output

23. Explain why the hardwired circuit in Figure 6 must be reconfigured when it is translated into a PLC ladder diagram.

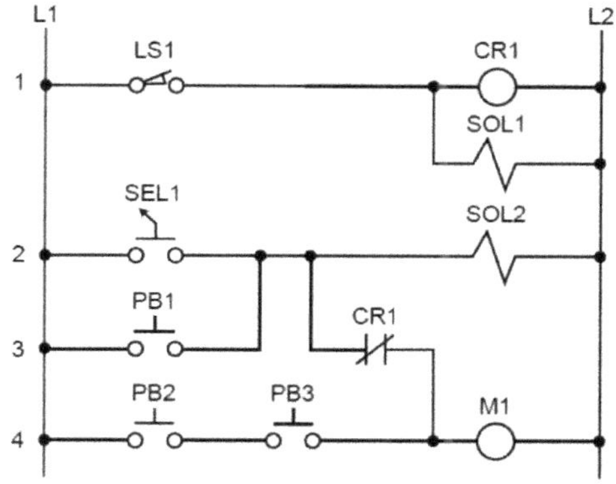

Figure 6. Circuit for problem 23.

24. Program _____ is the process of translating logic or relay contact diagrams into PLC ladder form.

25. Assuming that inputs use addresses 000–027, outputs use addresses 030–047, internals start at address 100_8, timers start at address 200_8, and an internal output is used to trap the instantaneous timer contacts, use the circuit shown in Figure 7 to:

(a). assign the internal addresses

(b). assign the I/O addresses

(c.) draw the I/O connection diagram

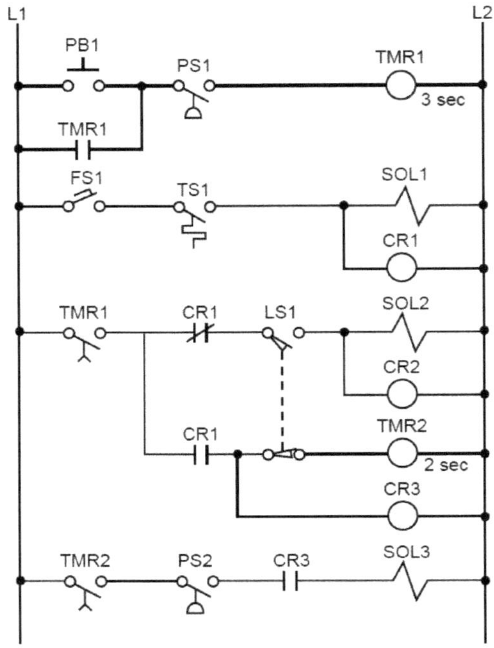

Figure 7. Circuit for problem 25.

26. Given that the stop push button will be wired as normally open, use the circuit in Figure 8 to:

(a). assign the I/O addresses

(b). draw the I/O connection diagram

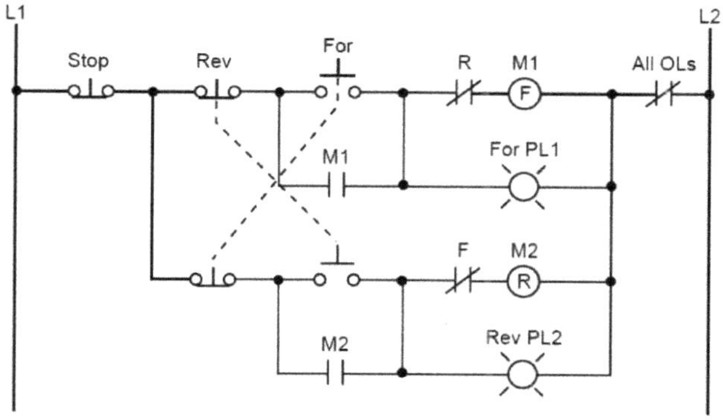

Figure 8. Circuit for problem 26.

27. Circle the locations where timer traps will be used in the PLC implementation of the circuit in Figure 9.

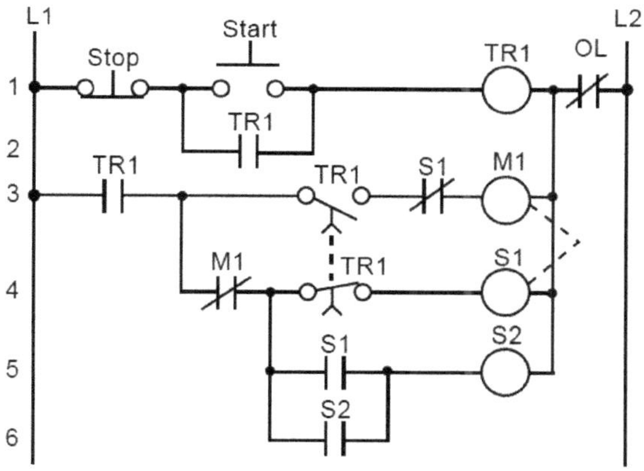

Figure 9. Circuit for problem 27.

28. Figure 10 shows a variable speed drive that is manually controlled by an operator station. What input field devices are required for the PLC implementation of this station?

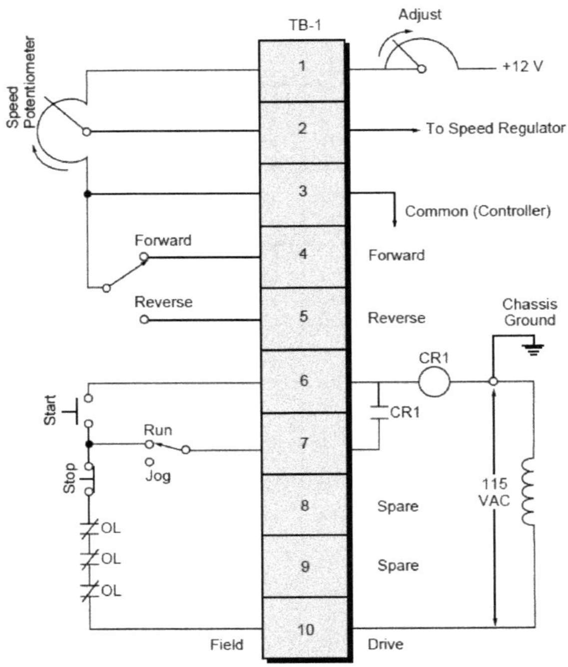

***Figure 10**. Variable speed drive.*

29. The PLC implementation of the large relay circuit in Figure 11 should include a normally open PLC fault contact and use internals to trap the timers. The PLC system has capacity for 512 I/O (000 to 777 octal). Inputs should start at address 000_8 and outputs should start at address 030_8. Internals should have addresses 1000–1777, MCRs should have addresses 2000–2037, and timers should have addresses 2040–2137. Using this large relay circuit:

(a). indicate the portions to be left hardwired

(b). assign the I/O addresses

(c). assign the internal addresses

(d). draw the I/O connection diagram

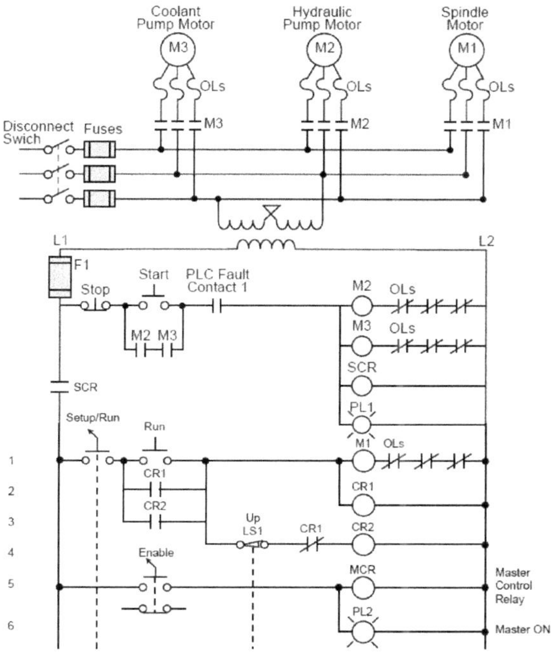

Figure 11a. Large relay circuit.

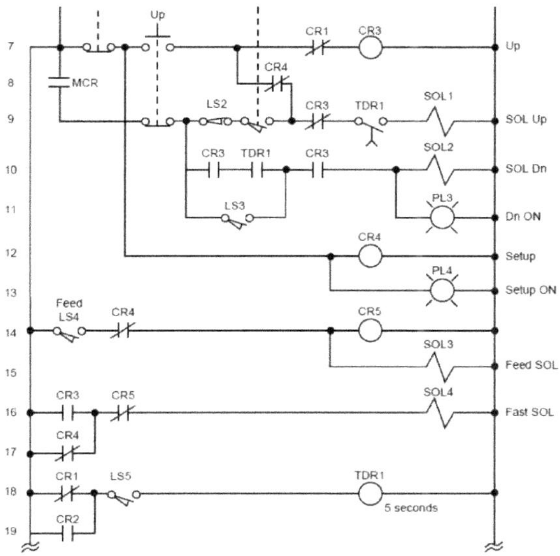

Figure 11b. Large relay circuit.

7.2. Solutions to the Problems

1. C.

2. algorithm

3. Guidelines for modernizing a control system include:

- understanding the actual process or machine function
- reviewing the machine logic and optimizing it when possible
- assigning real I/O addresses and internal addresses to inputs and outputs
- translating relay ladder diagrams into PLC coding

4. relay ladder diagram

5. B.

6. flowchart

7. D.

8.

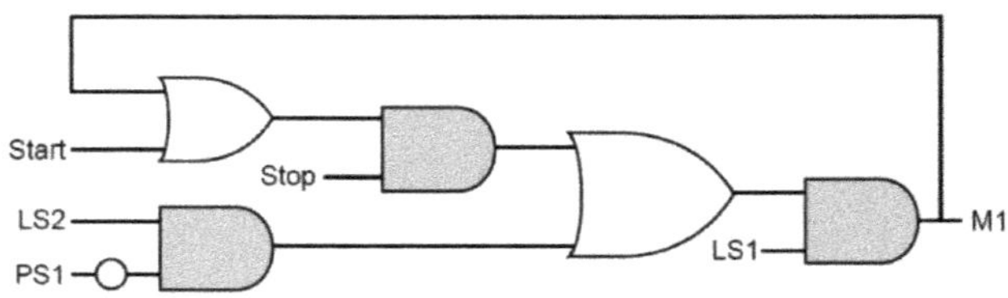

9.

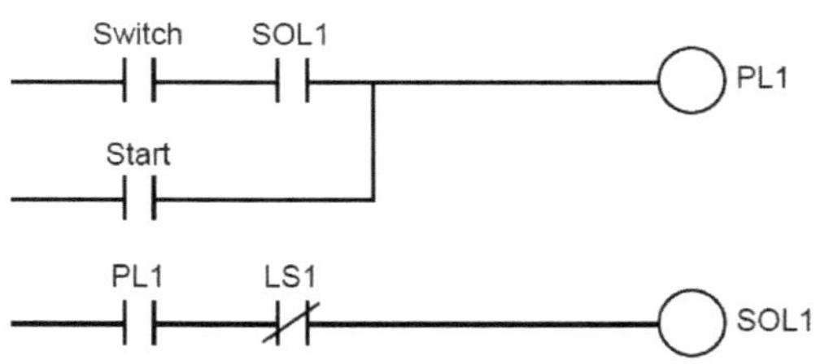

10. false; internal outputs are also documented during address assignment.

11. octal, decimal, hexadecimal.

12. I/O connection diagram.

13. (a).

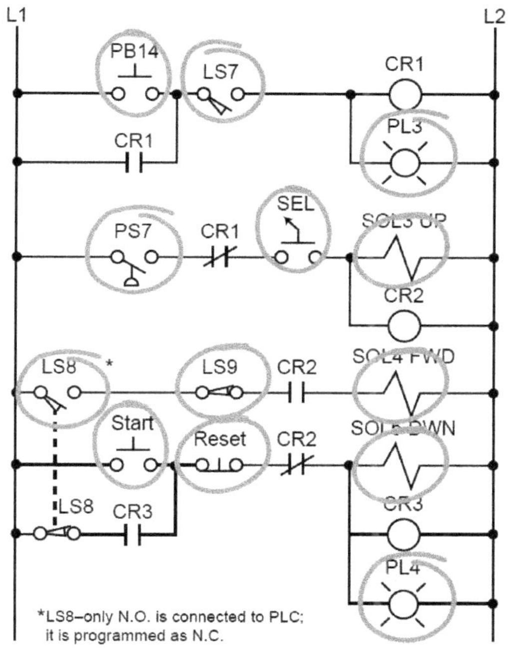

(b).

Module Type	Rack	Group	Terminal	Description
	I/O Address			
Input	0	0	0	PB14
	0	0	1	LS7
	0	0	2	PS7
	0	0	3	SEL
	0	0	4	LS8
	0	0	5	LS9
	0	0	6	Start PB
	0	0	7	Reset PB
Spare	0	1	0	Not used
	.	.	.	
	.	.	.	
	0	1	7	
Output	0	2	0	PL3
	0	2	1	SOL3 Up
	0	2	2	SOL4 Forward
	0	2	3	SOL5 Down
	0	2	4	PL4
	0	2	5	—
	0	2	6	—
	0	2	7	—

(c).

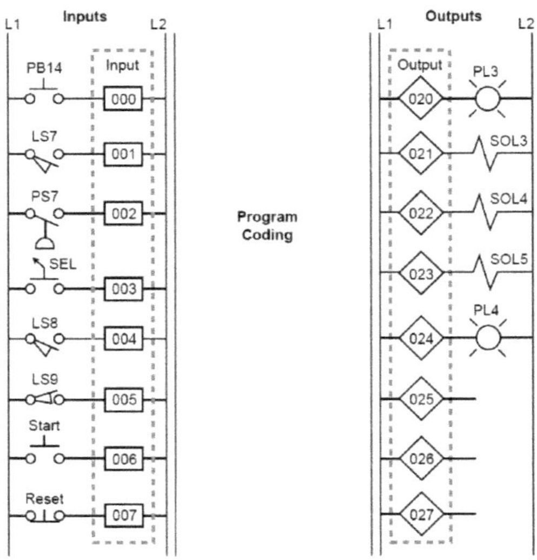

14. C.

15. B.

16. D.

17. A safety control relay is used to remove power from the I/O modules during a system error. When a malfunction occurs, the safety control relay will turn off, opening its SCR contact to stop the flow of power to the connected devices.

18. false; most of the time a normally closed input device is programmed as normally open; however, the programming of the input will depend on its function in the program

19. The normally closed PLC fault contacts are used to energize the PLC failure alarm. If the PLC fails, the PLC fault coil will not energize. Therefore, the normally open PLC fault contacts will not close to provide power to the connected devices. Instead, the normally closed PLC fault contacts will remain closed, sounding the PLC failure alarm.

20. (a).

Module Type	I/O Address: Rack	Group	Terminal	Description
Input	0	1	0	LS14
	0	1	1	SS3
	0	1	2	PS1
	0	1	3	S4
	0	1	4	LS15
	0	1	5	SS4
	0	1	6	—
	0	1	7	—
Output	0 ⋮	5 ⋮	0 ⋮	SOL7
Internal	1	0	0	CR10

(b).

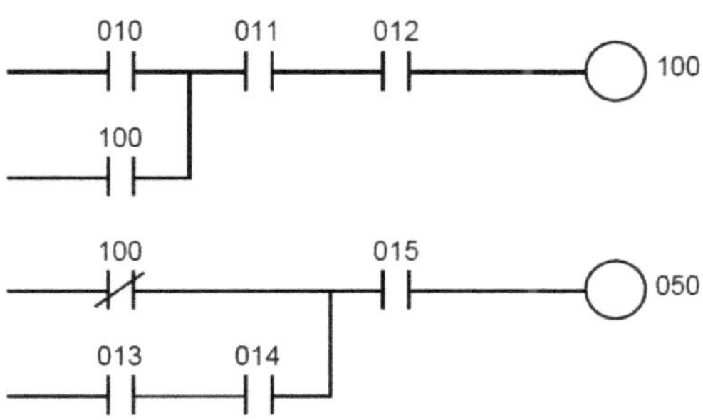

21. true

22. D.

23. There is a possibility of bidirectional power flow through the normally closed contact CR1 in line 3. A PLC will only allow power to flow in a forward path. Therefore, if the reverse path from line 4 to line 2 is meant to be followed, the circuit would have to be reconfigured so that the CR1 contacts are included in both lines 2 and 4.

24. coding

25. (a).

Device	Internal	Description
TMR1	100	Used to trap TMR1
CR1	—	Same as SOL1
CR2	—	Same as SOL2
CR3	101	Replace CR3
TMR1	200	Timer 1
TMR2	201	Timer 2

(b).

Module Type	I/O Address Rack	Group	Terminal	Description
Input	0	0	0	PB1
	0	0	1	PS1
	0	0	2	FS1
	0	0	3	TS1
	0	0	4	LS1
	0	0	5	PS2
	0	0	6	—
	0	0	7	—
Output	0	3	0	SOL1
	0	3	1	SOL2
	0	3	2	SOL3
	⋮	⋮	⋮	

(c).

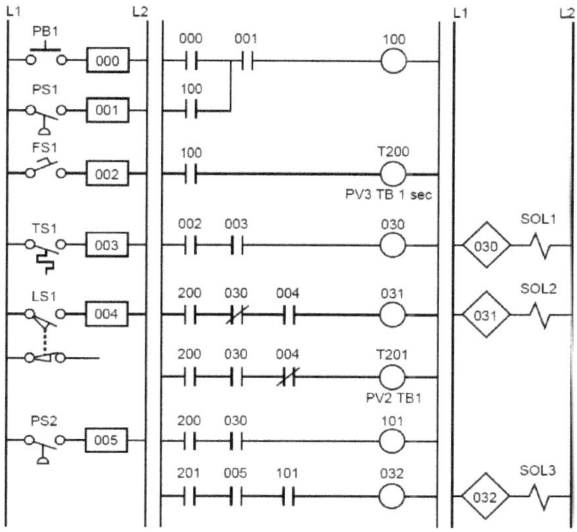

26. (a).

Module Type	I/O Address Rack	Group	Terminal	Description
Input	0	0	0	Stop PB (wired NC)
	0	0	1	Forward PB (wired NO)
	0	0	2	Reverse PB (wired NO)
	0	0	3	Overload contacts
Input	0	0	4	Acknowledge OL/Reset PB
	•	•	•	
	•	•	•	
	•	•	•	
Output	0	3	0	Motor starter M1 (FWD)
	0	3	1	Forward PL1
	0	3	2	Motor starter M2 (REV)
	0	3	3	Reverse PL2

(b).

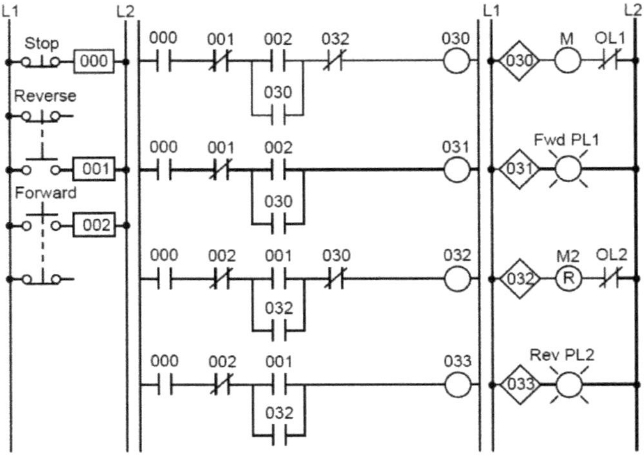

27.

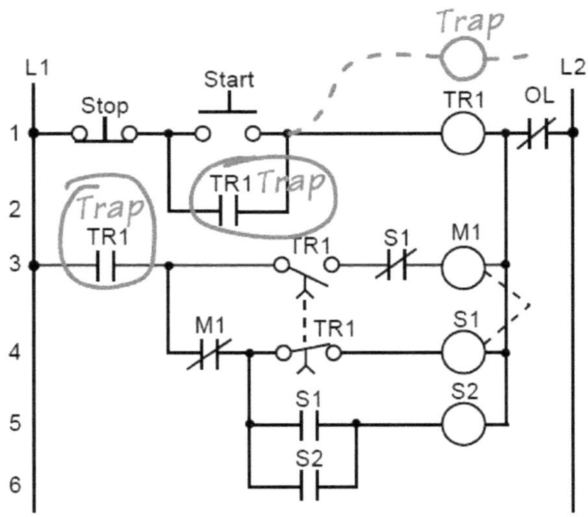

28. The field input devices are the start push button, stop push button, jog/run selector switch, and forward/reverse selector switch. The speed potentiometer will be replaced by an analog output in the PLC implementation.

29. (a).

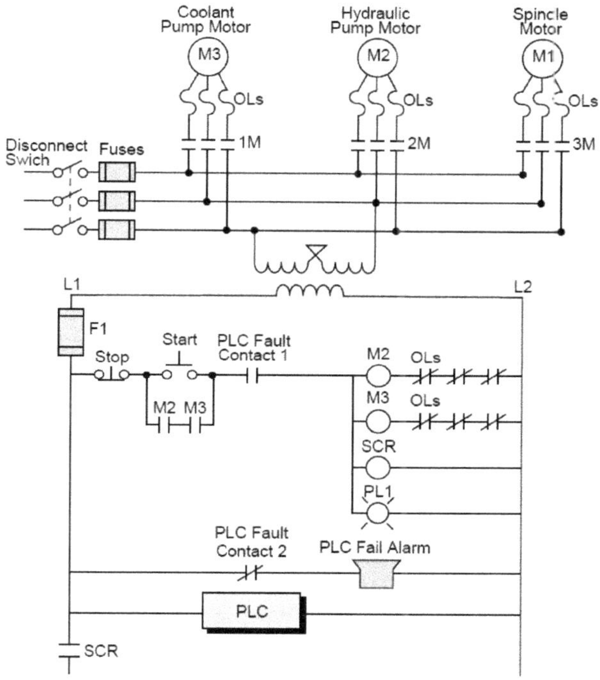
Coolant
Pump Motor
M3
Hydraulic
Pump Motor
M2
Spincle
Motor
M1
OLs
1M
2M
3M
Disconnect
Swich
Fuses
L1
L2
F1
Stop
Start
PLC Fault
Contact 1
M2 M3
M2
OLs
M3
OLs
SCR
PL1
PLC Fault
Contact 2
PLC Fail Alarm
PLC
SCR

(b).

Module Type	I/O Address Rack	Group	Terminal	Description
Input	0 0 0 0	0 0 0 0	0 1 2 3	Setup/Run Run PB Up LS1 Enable SS
Input	0 0 0 0	0 0 0 0	4 5 6 7	Up PB LS2 LS3 Feed LS4
Input	0 0 . . 0	1 1 . . 1	0 1 . . 7	LS5 Not used . .
Output	0 0 0 0	3 3 3 3	0 1 2 3	M1 Starter PL2 Master On SOL1 Up SOL2 Down
Output	0 0 0 0	3 3 3 3	4 5 6 7	PL3 Down On PL4 Setup On SOL3 Feed SOL Fast Feed

(c).

Device	Internal	Description
CR1	—	Same as M1
CR2	1000	Replace CR2
MCR	MCR 2000	First MCR address—replace MCR
CR3	1001	Replace CR3
CR4	—	Same as PL4
CR5	—	Same as SOL3
—	1002	Trap timer
TDR1	T 2040	First timer address—replace TDR1
MCR	END 2000	END MCR logic section
—	1003	Same as SOL1 in MCR fence

(d).

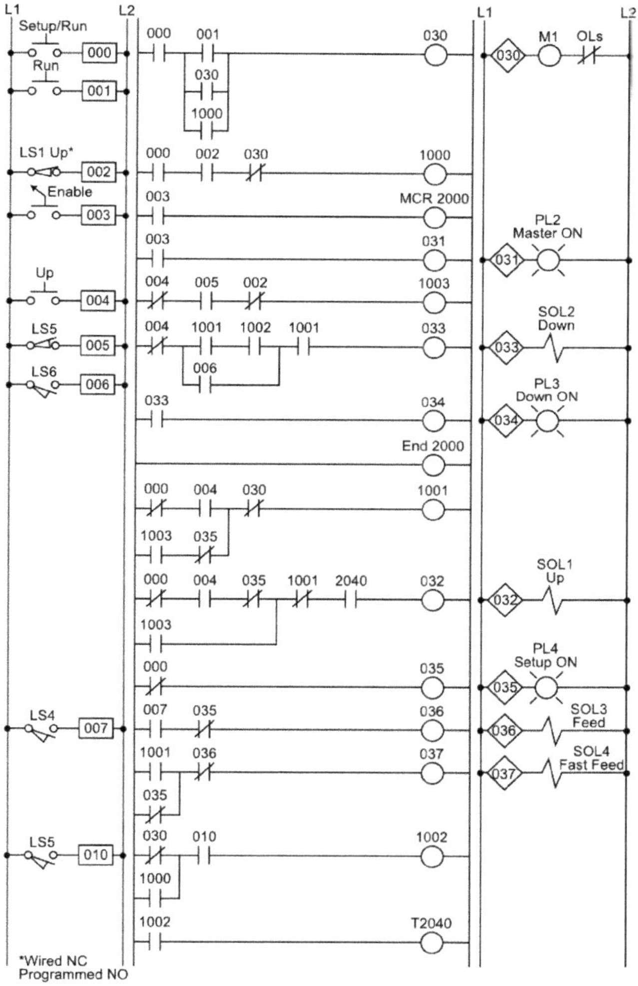
L1
L2
Setup/Run
Run
LS1 Up*
Enable
Up
LS5
LS6
LS4
LS5
*Wired NC
Programmed NO
MCR 2000
End 2000
T2040
M1
OLs
PL2
Master ON
SOL2
Down
PL3
Down ON
SOL1
Up
PL4
Setup ON
SOL3
Feed
SOL4
Fast Feed

8. More Helpful Resources

Here's the link to download all the graphics and images used in this book. You could find them helpful for your revision.

https://ln5.sync.com/dl/50226c8e0/nu2j2fp8-m5sa2t9c-h4hp6ix4-fbtebh56

Here's the Short BIT.LY link: https://bit.ly/3oXz1XZ

If you have questions or requests, or if you want to share your results with me, send them through to my email below.

Cheers,

Bolakale Aremu
Ojula Technology Innovations

OjulaTech@gmail.com

CPSIA information can be obtained
at www.ICGtesting.com
Printed in the USA
LVHW081907010323
740706LV00012B/1221

9 789493 689497